MAMMALS
of the
ROCKY MOUNTAINS

Chris Fisher

Do

Tama

Lone Pine Publishing

First printed in 2000 10 9 8 7 6 5 4 3 2
Printed in Canada

The Publisher: Lone Pine Publishing

10145 – 81 Ave.
Edmonton, AB T6E 1X5
Canada

1901 Raymond Ave. SW, Suite C
Renton, WA 98055
USA

202A, 1110 Seymour St.
Vancouver, BC V6B 3N3
Canada

Website: http://www.lonepinepublishing.com

Canadian Cataloguing in Publication Data

Fisher, Chris C. (Christopher Charles), 1970–
Mammals of the Rocky Mountains

Includes index.
ISBN 1-55105-211-3

1. Mammals—Rocky Mountains—Identification. I. Pattie, Donald L., 1933– II. Hartson, Tamara, 1974– III. Title.
QL719.R63F57 2000 599'.0978 C00-910416-X

Editorial Director: Nancy Foulds
Project Editor: Roland Lines
Editorial: Roland Lines, Eli MacLaren
Production Manager: Jody Reekie
Book Design: Heather Markham
Cover Design: Rob Weidemann
Layout & Production: Elliot Engley, Heather Markham
Cartography: Rob Weidemann, Curtis Pillipow, Elliot Engley
Cover Photograph: Thomas Kitchin/First Light
Separations & Film: Elite Lithographers Company

Photograph & Illustration Credits
All the photographs are by Terry Parker, except as follows: Michael H. Francis, p. 71; Leslie Degner, p. 95; Mark Degner, pp. 107 & 111; Wayne Lynch, pp. 115, 145 & 149.

All the animal illustrations are by Gary Ross, except the ones on pages 167, 176, 181, 182, 183, 205, 211, 223, 225, 229, 275, 279, 281 & 285, which are by Kindrie Grove. All the track illustrations are by Ian Sheldon.

We acknowledge the financial support of the Government of Canada through the Book Publishing Industry Development Program (BPIDP) for our publishing activities.

PC: P3

CONTENTS

REFERENCE GUIDE

HOOFED MAMMALS

American Bison
p. 28

Mountain Goat
p. 32

Bighorn Sheep
p. 36

Dall's Sheep
p. 40

Pronghorn
p. 44

Elk
p. 48

Mule Deer
p. 52

White-tailed Deer
p. 56

Moose
p. 60

Caribou
p. 64

Horse
p. 68

Mountain Lion
p. 74

Canada Lynx
p. 78

Bobcat
p. 82

Western Spotted Skunk
p. 86

CARNIVORES

Gray Wolf
p. 134

Red Fox
p. 138

Swift Fox
p. 142

Common Gray Fox
p. 146

RODENTS

Common
Porcupine, p. 152

Meadow Jumping
Mouse, p. 156

Western Jumping
Mouse, p. 157

Western Harvest
Mouse, p. 158

Deer Mouse
p. 160

Canyon Mouse
p. 162

Brush Mouse
p. 163

Pinyon Mouse
p. 164

Northern Rock
Mouse, p. 165

Northern Grasshopper
Mouse, p. 166

White-throated
Woodrat, p. 168

Mexican Woodrat
p. 169

Bushy-tailed Woodrat, p. 170

Norway Rat p. 172

House Mouse p. 174

Southern Red-backed Vole, p. 176

Western Heather Vole, p. 177

Water Vole p. 178

Meadow Vole p. 180

Montane Vole p. 181

Long-tailed Vole p. 182

Sagebrush Vole p. 183

Common Muskrat, p. 184

Brown Lemming, p. 186

Northern Bog Lemming, p. 187

American Beaver p. 188

Olive-backed Pocket Mouse, p. 192

Great Basin Pocket Mouse, p. 194

Silky Pocket Mouse p. 195

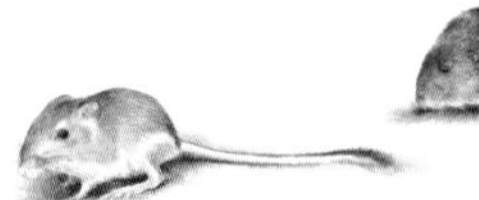
Ord's Kangaroo Rat, p. 196

Northern Pocket Gopher, p. 198

Idaho Pocket Gopher, p. 200

Botta's Pocket Gopher, p. 201

Least Chipmunk p. 202

Yellow-pine Chipmunk, p. 204

Cliff Chipmunk p. 205

Colorado Chipmunk p. 206

Hopi Chipmunk p. 208

Red-tailed Chipmunk, p. 209

Uinta Chipmunk p. 210

Woodchuck p. 212

Yellow-bellied Marmot, p. 214

Hoary Marmot p. 216

Columbian Ground Squirrel, p. 218

Richardson's Ground Squirrel, p. 220

RODENTS

Wyoming Ground Squirrel, p. 222

Uinta Ground Squirrel, p. 224

Thirteen-lined Ground Squirrel, p. 226

Idaho Ground Squirrel, p. 228

Rock Squirrel p. 229

Golden-mantled Ground Squirrel, p. 230

White-tailed Prairie-Dog, p. 232

Gunnison's Prairie-Dog, p. 234

Eastern Fox Squirrel p. 236

Albert's Squirrel p. 238

Red Squirrel p. 240

Northern Flying Squirrel, p. 242

HARES & PIKAS

Pygmy Rabbit p. 245

Mountain Cottontail, p. 246

Desert Cottontail p. 248

Snowshoe Hare p. 250

White-tailed Jackrabbit
p. 252

Black-tailed Jackrabbit
p. 254

American Pika
p. 256

Brazilian Free-tailed
Bat, p. 259

Big Free-tailed Bat
p. 260

Fringed Bat
p. 261

Long-eared Bat
p. 262

Northern Bat
p. 263

California Bat
p. 264

Western Small-footed
Bat, p. 265

Little Brown Bat
p. 266

Yuma Bat
p. 268

Long-legged Bat
p. 269

Hoary Bat
p. 270

Silver-haired Bat
p. 272

Big Brown Bat
p. 273

Spotted Bat
p. 274

Pallid Bat
p. 275

Townsend's Big-eared Bat, p. 276

Masked Shrew
p. 279

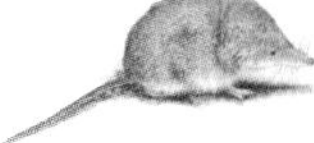

Preble's Shrew
p. 280

Vagrant Shrew
p. 281

Dusky Shrew
p. 282

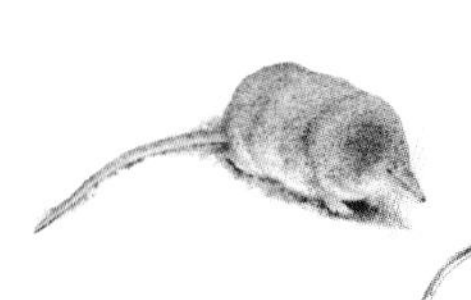

Dwarf Shrew
p. 283

Common Water Shrew
p. 284

Arctic Shrew
p. 286

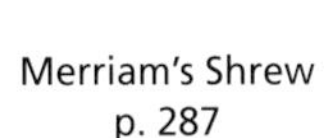

Merriam's Shrew
p. 287

Pygmy Shrew
p. 288

Desert Shrew
p. 289

Virginia Opossum
p. 290

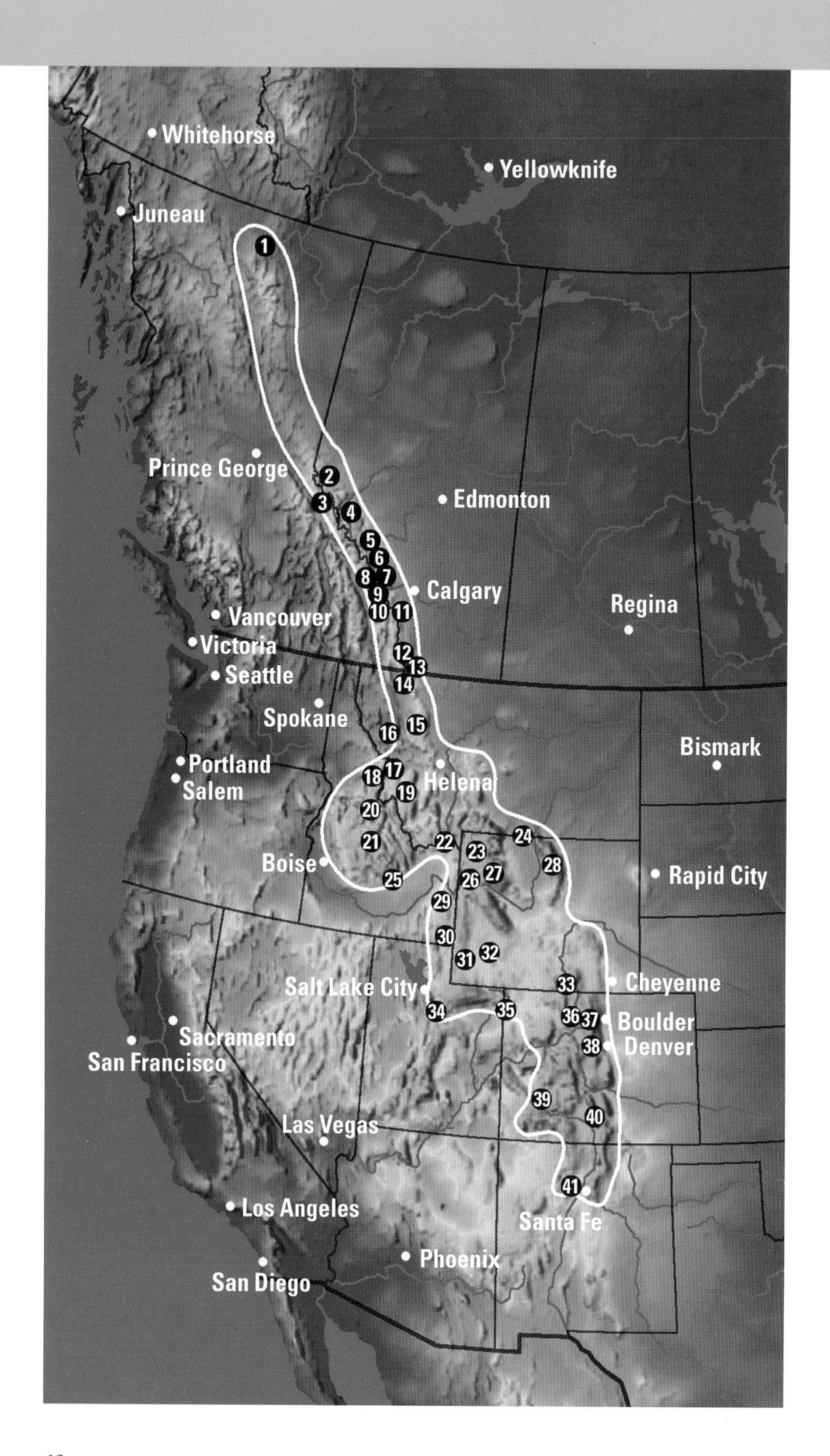
Whitehorse
Yellowknife
Juneau
1
Prince George
2
3
4
Edmonton
5
6
8
7
9
Calgary
10
11
Regina
Vancouver
Victoria
12
13
Seattle
14
Spokane
15
16
Bismark
Portland
Salem
18
17
Helena
19
20
21
22
23
24
28
Boise
25
26
27
Rapid City
29
30
31
32
Salt Lake City
33
Cheyenne
34
35
36
37
Boulder
38
Denver
Sacramento
San Francisco
39
40
Las Vegas
41
Santa Fe
Los Angeles
Phoenix
San Diego

Introduction

Few things characterize wilderness as well as wild animals, and few animals are more recognizable than our fellow mammals. In fact, many people use the term "animal" when they really mean "mammal"—they forget that birds, reptiles, amphibians, fish and all the many kinds of invertebrates are animals, too.

Mammals come in a wide variety of colors, shapes and sizes, but they all share two characteristics that distinguish them from the other vertebrates: only mammals have real hair, and only mammals nurse their young from mammary glands (the feature that gives this group its name). Other, less well-known features that are unique to mammals include a muscular diaphragm, which separates the lower abdominal cavity from the cavity that contains the heart and lungs, and a lower jaw that is composed of a single bone on each side. Additionally, a mammal's skull joins with the first vertebra at two points of contact—a bird's or reptile's skull only has one point of contact, which is what allows birds to turn their heads so far around. As well as setting mammals apart from all other kinds of life, these characteristics also identify humans as part of the mammalian group.

The Rocky Mountains provide some of the best mammal-watching opportunities in North America, whether you are walking among ground squirrels in a meadow, watching a beaver swim in the evening light or experiencing the haunting sounds of bugling Elk. Much has changed over the last 150 years, but the Rocky Mountains remain as an internationally recognized destination for visitors who are interested in rewarding natural experiences. To honor this treasure is to celebrate North America's intrinsic virtues, and this book is intended to provide readers with the knowledge needed to build an appreciation of the rich variety of mammals in the mountains. Whether you are a naturalist, a photographer, a wildlife enthusiast or all three, you will find terrific opportunities in the Rockies to satisfy your greatest wilderness expectations.

Mammal-Watching Sites in the Rocky Mountains

1 The Muskwa
2 Willmore Wilderness Park
3 Mount Robson & Rearguard Falls Provincial Parks
4 Jasper National Park
5 White Goat Wilderness Area
6 Siffleur Wilderness Area
7 Banff National Park
8 Yoho National Park
9 Mount Assiniboine Provincial Park
10 Kootenay National Park
11 Kananaskis Country & Peter Lougheed Provincial Park
12 Castle Crown Wilderness
13 Waterton Lakes National Park
14 Glacier National Park
15 Great Bear, Bob Marshall & Scapegoat Wilderness Areas
16 National Bison Range
17 Bitterroot Watchable Wildlife Triangle
18 Selway-Bitterroot Wilderness Area
19 Anaconda-Pintler Wilderness
20 Frank Church–River of No Return Wilderness
21 Sawtooth National Recreation Area, Sawtooth Wilderness Area & Land of the Yankee Fork State Park
22 Red Rock Lakes National Wildlife Refuge
23 Yellowstone National Park
24 Pryor Mountain Wild Horse Range
25 Craters of the Moon National Monument
26 Grand Teton National Park
27 Teton & Gros Ventre Wildernesses
28 Cloud Peak Wilderness
29 Grays Lake National Wildlife Refuge
30 Bear Lake National Wildlife Refuge
31 Fossil Butte National Monument
32 Seedskadee National Wildlife Refuge
33 Encampment Wildernesses
34 Timpanogos Cave National Monument
35 Dinosaur National Monument
36 Arapaho National Wildlife Refuge & Illinois River Moose Viewing Site
37 Rocky Mountain National Park
38 Golden Gate Canyon State Park
39 Big Blue & Powderhorn Wilderness Areas
40 Great Sand Dunes National Monument
41 Bandelier National Monument

Rocky Mountain Geology

The Rocky Mountains are a dominant landscape feature of North America, forming the continent's backbone—the Continental Divide. This height of land separates waters flowing to the Pacific Ocean from waters flowing to the Arctic Ocean or the Gulf of Mexico. Some of North America's largest rivers began as glaciers or small streams in the Rocky Mountains. This chain of mountains is more than 300 mi (480 km) wide in places and stretches more than 3000 mi (4800 km) from northern British Columbia to the Sangre de Cristo Mountains of New Mexico.

The Rockies are commonly divided into four segments: the Canadian Rockies (which geologically extend into northern Montana), the Northern U.S. Rockies (in Montana and central Idaho), the Central U.S. Rockies (mostly in Wyoming and northern Utah) and the Southern U.S. Rockies (in southeastern Wyoming, Colorado and northern New Mexico. The Wyoming Basin, which sits between the Central and Southern U.S. Rockies, is sometimes considered a fifth segement. This gap in the Rockies resulted from whole mountain ranges being buried in sediment, and it now acts as a peninsula of the Great Plains.

The geological history of the Rockies is a fascinating and complex story that scientists are still piecing together—it is estimated to span no less than 1.5 billion years. Geological study indicates that different parts of the Rockies were formed at different times and in different ways. Vertical thrusting and faulting of sedimentary rocks that were once the bed of an ancient sea formed some of the Rocky Mountain ranges. Some ranges are made of metamorphic rocks that formed underground under tremendous heat and pressure before becoming exposed to the surface; other ranges are composed of igneous rocks created by volcanic processes.

More recently, continental glaciers sculpted many of these mountain ranges into the deep, U-shaped valleys, steep slopes and high peaks we see today. The remnants of these glaciers still cover many of the mountain crests in the Canadian and Northern U.S. Rockies. The highest mountain peaks in the Rocky Mountains are found in the Southern U.S. segment, with 54 summits reaching over 14,000 ft (4270 m) above sea level. As you can imagine, each of the segments of the Rockies displays unique scenic and biological characteristics based on its geological history.

Elk

Life Zones

The Rocky Mountains contain different growing environments, known as "biological zones" or "life zones." Life zones are the product of unique combinations of geology, climate, elevation, latitude, slope direction (aspect) and slope angle. The following section lists these zones and the flora and fauna found within them.

Alpine (Tundra)

The alpine zone comprises the bare rocks, glaciers, tundra and meadows above treeline. This cold, windswept environment may have snow-free areas early in spring and even through most of winter, but the alpine may also lay blanketed with drifts for most or all of summer. The temperature is just sufficient for enough days to permit vegetative growth. At treeline, the ground is thawed long enough for the trees to gain their yearly supply of soil moisture and minerals, and summer growth barely replaces needles and twigs killed in the winter. Fewer species of plants and animals survive in the alpine than in other ecoregions.

The high peaks of the Rocky Mountains are among the highest in the world and they call out to climbers. The highest peak in the Rockies is Mount Elbert, in Colorado, at 14,433 ft (4399 m), with an ascent of 2400 ft (731.5 m) from the base to the peak. Mount Robson, in British Columbia, is 12,972 ft (3954 m), but it has a more significant ascent (9741 ft [2969 m] from base to peak).

Subalpine

The subalpine zone exists upward from the upper edge of the montane to treeline, and it consists of dense clumps of evergreens and wildflower meadows. Mid-elevation slopes have heavy forests with cool, damp, mossy forest floors and receive the heaviest snow accumulation. Farther upslope, the trees become shorter or shrub-like, often on the eastern, leeward side of the rocks and ridges, where the trees are more sheltered under the snow and can therefore escape winds and winter storms. These trees are often called "kruppelholz" or "krummholz." The subalpine is what most people envision when they think of the Rocky Mountains. Stunning waterfalls, rocky cliffs dotted with Mountain Goats and Golden Eagles soaring against an impressive backdrop of towering mountain peaks are a few of the spectacular images this ecoregion offers. Travelers often visit the subalpine's ski resorts and secluded cabins to add adventure and romance to their mountain vacation.

Montane

The montane zone consists of the lower slopes and valleys above the foothills. The western slopes are wetter, heavier and shrubbier compared to the drier eastern slopes. The montane also holds the greatest variety of flowers, trees and shrubs. Montane valleys are critical winter habitat for Elk, deer and Bighorn Sheep, and therefore also for their predators, such as the Gray Wolf and the Coyote. This habitat is essential to wildlife, but it is small (only 2 to 10 percent of the Rockies) and it is the same part of the Rockies favored by humans. It is in the montane that the heaviest effects of human encroachment occur, making it a threatened ecoregion in need of protection and restoration.

Foothills

The foothills are the transition from either the boreal forest or the plains and prairies to the mountains. It is a common misconception that the foothills are part of the mountains. The upthrusting event that formed the Rocky Mountains also created the foothills to the east by causing buckles or ripples in the bedrock, but this rippled land is geologically and biologically distinct. The first low-elevation slopes before the treed montane slopes are considered the foothills. In the U.S. the foothills are generally low-elevation scrublands that blend into the plains. The cooler, moister, north-facing slopes and valleys are where shrubs first grow, but with increasing elevation they spread to south-facing slopes. Trees then begin to appear on the northern slopes and valleys and eventually they become the montane forests. Grasses provide scattered ground cover in these dry communities. In northern Canada the transition is less distinct: the foothills blend into the boreal forest in the east and the Columbian forest in the west.

Plains and Prairies

This zone is the only one that is technically not part of the Rocky Mountains. The plains and prairies lie in the rainshadow east of the Rockies and blend into the foothills but do not form a definite zone within the mountains. Scattered tracts of grasslands can be found almost anywhere in the Rockies south of the Athabasca Valley in Jasper National Park.

Many mammals that were once widespread across the plains, such as the Grizzly Bear, American Bison, Elk, Gray Wolf and Mountain Lion, were exterminated by Euroamerican settlers.

Human-altered Landscapes

Many of the most common plants and animals found along roads, in fields and around townsites did not occur in the Rockies before to modern human habitation and transportation. House Mice, Norway Rats and Black Rats are some of the highly successful exotics that were introduced to North America from Europe and Asia.

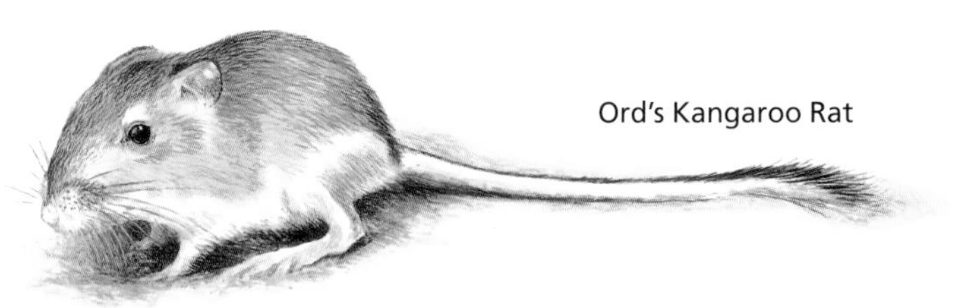

Ord's Kangaroo Rat

Seasonality

The extreme seasons of the Rocky Mountains have considerable influences on the lives of mammals. With the exception of bats, Rocky Mountain mammals are confined to relatively slow forms of terrestrial travel. As a result, they have limited geographic ranges, and they must cope in various ways with the changing seasons.

With the receding of winter snows and the greening of the mountain region, spring signals renewal. It is at this time of year that many mammals bear their young. The lush new growth provides ample food for herbivores, and with the arrival of new herbivore young, the predatory mammals enjoy good times as well. While some small mammals, particularly the shrews and rodents, mature within weeks, the offspring of the larger mammals are dependent on their parents for much longer periods.

During the warmest time of the year, the animals' bodies have recovered from the strains of the previous winter's food scarcity and spring's reproductive efforts, but summer is not a time of relaxation. To prepare yet again for the upcoming fall and winter, some animals must eat vast quantities of food to build up fat reserves, while others work furiously to stockpile food caches in safe places. Some mammals, such as the Richardson's Ground Squirrel, start hibernation as early as late July, which signals to the keen naturalist that the slow slope to winter has already begun.

For some of the more charismatic species, fall is the time for mating. At this time of the year, the bugling bull Elk demonstrate extremes in aggression and vigilance. Some small mammals, however, such as voles and mice, mate every few months, and the last of the year's litters are often just starting out on their own as the first snows begin to dust the land. These small rodents must adjust quickly if they are to survive the fast-approaching winter.

Winter is the most difficult time for many Rocky Mountain mammals. For many herbivores, high-energy foods are difficult to find, often requiring more energy to locate than they provide in return. This food deficit gradually weakens most mammals through winter, and the ones that were not fit enough at the start of winter end up feeding the equally needy carnivores, which ironically find an ally in winter's severity. Voles and mice do not necessarily share this viewpoint—an insulating layer of snow buffers their elaborate trails from the worst of winter's cold. Food, shelter and warmth are all found in this thin layer, and the months devoted to food storage now pay off. Winter eventually wanes, and while death is an omnipresent associate, this season sets the foundation for another springing of life.

An important aspect of seasonality is its effect on species composition. When you visit the mountains in winter, you will see a different group of mammals than in the summer. Many species, such as most squirrels and the charismatic bears, are dormant in winter. Conversely, many ungulates may be more visible in winter because they enter lowland meadows to find edible vegetation.

Watching Mammals

Many types of mammals are most active at night, so the best times for viewing mammals are during the "wildlife hours" at dawn and at dusk. At these times of day, mammals are out from their daytime hideouts, moving through areas where they are more easily encountered. During winter, hunger may force certain mammals to be more active during midday. Conversely, when the conditions are more favorable during spring and summer, mammals may become less active.

While many of the mammals that are encountered in the mountains often appear easy to approach, it is important to respect your own safety as much as the safety of the animal being viewed. This advice seems obvious for the larger species (although it is ignorantly dismissed in some instances), but small mammals should also be treated with respect. Honor both the encounter and the animal by demonstrating a respect that is owed to the special occasion.

With the Rocky Mountain's abundance of large protected areas, many of the larger mammals, in particular, can be easily viewed from the safety of a vehicle along the many roadways that cut through the parks. If you walk the backcountry trails, however, you can find yourself in the very homes of some mammals.

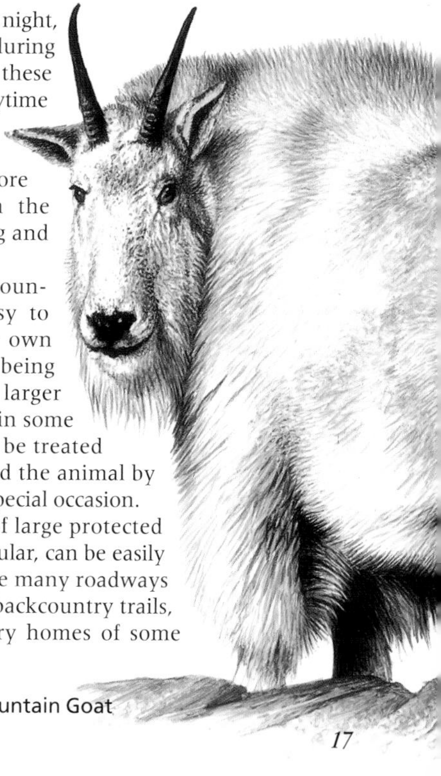

Mountain Goat

The Rockies' Top Mammal-Watching Sites

BRITISH COLUMBIA

The Muskwa

In the fall of 1997, the government of British Columbia announced the protection of 2.5 million acres (1 million hectares) of wilderness in the northern Rockies. This massive park is the first major contribution to the Yellowstone to Yukon initiative, which seeks to create a series of large protected areas and buffer zones along the Rocky Mountains.

Mount Robson & Rearguard Falls Provincial Parks

At 12,972 ft (3954 m), Mount Robson is the highest peak in the Canadian Rockies. Berg Glacier oozes over its north face, occasionally creating a thunderclap as massive chunks of ice calve into Berg Lake. Backcountry trails link hikers to Jasper National Park and typical high-country wildlife, while the marshes of Moose Lake give highway travelers a chance to see Moose and many species of waterfowl. At Rearguard Falls, watch for Chinook Salmon jumping the final barrier to reach their spawning grounds after an 800-mi (1300-km) journey from the Pacific.

Yoho National Park

The melting Daly Glacier, nestled among the towering peaks of the Continental Divide, gives rise to the spectacular Takakkaw Falls, whose water plummets 1248 ft (380 m) into the wild Yoho River. Grizzly Bears, Mountain Lions, Hoary Marmots and Mountain Goats are regular inhabitants of this park. Moose, deer, and Elk may be found at the salt lick near the amazing natural bridge. Yoho's Burgess Shale World Heritage Site contains the fascinating fossil remains of marine animals estimated to be 530 million years old.

Mount Assiniboine Provincial Park

Renowned as Canada's Matterhorn, Mount Assiniboine may look similar to its European counterpart but the communities of plants and animals found here are quite different from those of the Swiss Alps. Far from the drone of traffic and urban chaos, this park is only accessible by foot, horse or, in winter, snowshoe or ski. Adventurous and observant visitors might just meet Bighorn Sheep and hear the howling of wolves.

Kootenay National Park

Kootenay's ochre-tinted paint pots and myriad wild plants were once used by the Kootenai Nation for ceremonial and survival needs. Black huckleberry, wild strawberry, and yellow glacier-lily were gathered, while Mule Deer, Snowshoe Hares and American Beavers were hunted. In more recent times, this magnificent park's natural hot springs and many natural salt licks have made this a special place for tourists.

ALBERTA

Willmore Wilderness Park

In this rugged, remote, mountainous park, wild processes continue largely unimpaired by humans, allowing wild species to interact with each other as they have for centuries. It is a great place to see many large mammals—watch for Grizzly Bears and Gray Wolves—although you will find that the Mountain Goats and Bighorn Sheep are much more timid here than in nearby Jasper National Park.

Jasper National Park

Grizzly Bears, Black Bears, Gray Wolves, Caribou, Elk, Mule Deer, Moose, Mountain Lions and Bighorn Sheep all roam the wilds of this large park. Look for sheep and wolves at the north end of Jasper Lake; Caribou, American Mink and Moose along Medicine River; and bears along the Icefields Parkway. Visitors can hike on the toe of the Athabasca Glacier and learn about geological processes first-hand.

White Goat & Siffleur Wilderness Areas

These small, rugged front-range wilderness areas protect a wide variety of mammals and their habitats. Summer dayhikes or multi-day backcountry trips are sure to reveal Least Chipmunks and Mule Deer. Lucky observers may get a good look at a Common Porcupine, Hoary Bat, American Marten or Canada Lynx.

Banff National Park

Canada's first national park is magnificent and accessible. A wide range of enjoyable excursions provide world-renowned scenery and excellent chances to meet stunning flora and fauna. Lake Louise, Peyto Lake and Moraine Lake are three of the jewels of Banff National Park. The looping alpine meadow trails at Bow Summit are accessible to everyone and provide visitors with good chances to see Golden-mantled Ground Squirrels and Bighorn Sheep.

Kananaskis Country & Peter Lougheed Provincial Park

Winter in the front ranges of the Rockies is magical, revealing the tracks of Mountain Lions, Gray Wolves, Snowshoe Hares and Elk. In spring and summer, the forests and mountain slopes echo with the calls of ravens and the sweet voices of songbirds. Dippers dive in icy waters for caddisfly larva, while owls scan the flowery forest floor in search of voles and shrews.

Castle Crown Wilderness

Bordering the northern edge of Waterton Lakes National Park, this relatively unknown wilderness is unprotected from human development. The front range mountains and canyons, lush backcountry valleys, clean headwater streams and jagged peaks of the Continental Divide provide critical habitat for many endangered and threatened plant and animal species including Gray Wolves and Grizzly Bears.

Virginia Opossum

Waterton Lakes National Park

Forming the Canadian part of the Waterton-Glacier International Peace Park, Waterton Lakes National Park, located where the prairies meet the mountains, contains a unique mix of grasslands, forests, alpine tundra and vital aquatic habitats. The variety of habitats results in tremendous species diversity, including beavers, pikas, goats and sheep.

MONTANA

Glacier National Park

The Going-to-the-Sun Road (closed in winter) is one of Montana's most spectacular drives, with many roadside turnouts from which to enjoy the scenery. Although humans tend to be the most abundant species in summer, Mountain Goats, Bighorn Sheep, Mule Deer and bears are regularly near roadways. Lake McDonald and its old-growth shores provide excellent habitat for both aquatic and forest-dwelling animals.

Great Bear, Bob Marshall & Scapegoat Wilderness Areas

Straddling the Continental Divide, these great wilderness areas are administered by the U.S. Forest Service. They are vital links in the chain of mountain wilderness extending south from the Waterton-Glacier International Peace Park. Precipitous, rocky peaks, colorful alpine meadows, and rich, forested valleys harbor many interesting wild plants and animals. Forestry roads and state highways provide access to the campgrounds and trailheads on the perimeter of these areas.

National Bison Range

On the Flathead Indian Reservation, at the southern end of the Flathead Valley, the legacy of the American Bison continues, even though these mighty creatures no longer enjoy the numbers and freedom they once had. Elk, Mule Deer, White-tailed Deer, Bighorn Sheep and Pronghorn share the fenced valley-bottom range with up to 500 bison. The two-hour Red Sleep Mountain self-guided drive gives a close-up view of the bison and the other mammals and birds recorded here. Pablo and Ninepipe national wildlife refuges are also on the reservation, north of the National Bison Range.

Bitterroot Watchable Wildlife Triangle

The 60-mi (96-km) Bitterroot Valley south of Missoula offers a first-hand experience with Rocky Mountain nature. The Charles Waters Nature Trail, Lee Metcalf National Wildlife Refuge and Willoughby Environmental Education Area, all near Stevensville, provide nature trails and wildlife viewing blinds in excellent wildlife habitat. Otters, muskrats, porcupines, deer and a whole host of songbirds can be observed.

Anaconda-Pintler Wilderness

The 280 mi (450 km) of trails that wind through deep, U-shaped valleys, over glacial moraines and around cirque lakes can take you from 5100 ft (1550 m) to 10,790 ft (3290 m) in elevation. Moose thrive in the valley bottoms, while Mountain Goats are common in the high country. Black Bears, Grizzly Bears and Hoary Marmots dream away the cold mountain winters safe in their dens, while Bobcats, Coyotes and Elk leave a record of their travels on snowy ground.

Red Rock Lakes National Wildlife Refuge

This refuge, just west of Yellowstone National Park, protects important nesting habitat for swans, goldeneyes and cranes. Although birds are this refuge's main attraction, a quiet canoe trip is a good way to observe water-dwelling mammals, including muskrats, beavers and water shrews.

Pryor Mountain Wild Horse Range

The dramatic Pryor Mountains soar above the high plains near the Montana-Wyoming border, just west of the Bighorn Canyon. A herd of about 100 wild mustangs roam this range, and some people believe their lineage dates back to the 1700s. The Pryors are also a good place to see Bighorn Sheep and Pronghorns.

IDAHO

Selway-Bitterroot Wilderness Area

Hundreds of miles of maintained trails give intrepid visitors to northeastern Idaho non-motorized access to cool, old-growth Douglas-fir forests, clear mountain streams, flowery meadows and rocky slopes. Picnic areas and campgrounds are on the perimeter of the wilderness for those unable to make the commitment to backcountry pursuits. Least Chipmunks and White-tailed Deer add magic to the towering trees and growing plants of both front-country and backcountry campgrounds.

Frank Church–River of No Return Wilderness

Acclaimed as the largest wilderness area in the lower 48 states, this place of whitewater rivers, thriving natural forest communities and majestic mountains is especially known for its large population of secretive Mountain Lions. American Martens, Fishers, Red Foxes and Wolverines also thrive among healthy prey populations of small mammals, fish, reptiles and amphibians.

Sawtooth National Recreation Area, Sawtooth Wilderness Area & Land of the Yankee Fork State Park

The Sawtooth, Boulder and White Cloud Mountains cradle over 300 alpine lakes and form the headwaters of five major rivers. Lodges, guest ranches, campgrounds, picnic areas and dayhike trails allow you to spend some quality time investigating the flora and fauna while taking in the local culture. If you are more adventurous, backcountry trails, campsites and rafting tours enable you to immerse yourself in nature.

Craters of the Moon National Monument

Situated at the southern base of Idaho's Pioneer Mountains, this national monument protects an astonishing biotic community in a harsh volcanic, desert environment. Hardened lava tubes, spatter cones, cinder cones and lava flows are some of the incredible volcanic features that may be seen. Snowshoeing, cross-country skiing, animal tracking and photography are popular winter activities.

Grays Lake National Wildlife Refuge

Established to protect breeding cranes and waterfowl, this lake may produce up to 5000 ducks, 2000 geese and 150 sandhill cranes in a single breeding season. Although birds are the major attraction here, you have a good chance of seeing aquatic mammals from the wildlife viewing platform next to the ranger station. Most of the area is sensitive to human disturbance and is therefore closed to access.

Bear Lake National Wildlife Refuge

Straddling the Idaho-Utah border, Bear Lake's turquoise waters are surrounded by marsh and grassland vegetation that provide staging and breeding habitat for fascinating wildlife species. Many birds nest here during summer, and during winter Coyotes hunt for small rodents living under the snow.

WYOMING

Yellowstone National Park

The largest park in the lower 48 states brings volcanic natural forces alive with 10,000 exploding, bubbling, and steaming geothermal geysers, hot springs, steam vents and mud-pots. Yellowstone is also blessed with an incredible diversity of colorful flowers, trees and shrubs, and dynamic communities of wild animals. Wolves, bears, Moose, Elk and American Bison, among other fascinating mammals, will leave you breathless.

Grand Teton National Park

The Grand Tetons and Jackson Hole rank among the oldest formations in the Rocky Mountain system. The 42-mi (68-km) Teton Park Scenic Loop Drive is by far the most popular way of reaching the scenic viewpoints and wildlife viewing hotspots. From November through May, approximately 7500 Elk converge on the National Elk Refuge, which also supports Long-tailed Weasels, American Badgers and Uinta Ground Squirrels, among other animals.

Teton & Gros Ventre Wildernesses

Connecting the high mountains of the Continental Divide in Bridger-Teton National Forest to those in Grand Teton National Park, these wilderness areas provide prime habitat for Grizzly Bears, Bighorn Sheep, Mountain Lions and Fishers. Recreationalists should use minimal impact techniques to reduce unnecessary disturbances.

Cloud Peak Wilderness

Established along the crest of the Bighorn Mountains, this wild place offers 150 mi (240 km) of maintained trails. American Pikas, Bighorn Sheep and Yellow-bellied Marmots are just a few of the many mammals to observe and appreciate here.

Desert Cottontail

Fossil Butte National Monument

This monument protects the remnants of Wyoming's ancient past. Crocodiles, giant turtles, palms, cypress trees and long-extinct fish once lived in a large lake estimated to have existed here about 50 million years ago. Now this area's dry, high-desert habitats support Coyotes and White-tailed Jackrabbits, while wetter areas are home to Moose and Golden-mantled Ground Squirrels. Taking fossils from the park and camping overnight are not permitted, but hiking, photography and nature appreciation are encouraged.

Seedskadee National Wildlife Refuge

For early settlers, this flattened region represented the easiest route for crossing the near-impenetrable height of land running from Canada to New Mexico. The wildlife refuge was established in exchange for flooding the Green River to create the Flaming Gorge Reservoir, and it is a haven for beavers, ground squirrels, deer and other mammals relying on grassland, sagebrush and riparian habitat.

Encampment Wildernesses

Huston Park Wilderness, Savage Run Wilderness, Platte River Wilderness, and the Encampment River Wilderness Area are all small protected areas established by the U.S. Forest Service within the Medicine Bow National Forest along the Colorado border. The Continental Divide National Scenic Trail passes through this area, and it is a good place for hikers to see Long-eared Bats.

UTAH

Timpanogos Cave National Monument

This monument was established to preserve the colorful limestone cavern on the side of Mount Timpanogos. Open to visitors from Memorial Day to Labor Day, a three-hour hike and tour of the cave gives you a memorable view of the cave's spectacular helietites (formations that are created by water) growing in all directions throughout the cave. Watch for bats and mule deer.

Dinosaur National Monument

Flowing through deep canyons at the base of the Uinta Mountains, the waters of the Yampa River and the Green River slowly erode the ancient river valley to expose dinosaur fossils from the Jurassic Period. Dinosaur National Monument is one of few places in the Rockies where prehistoric fossils and aboriginal pictographs and petroglyphs exist alongside natural communities to give us a keener insight into a history that spans millions of years.

COLORADO

Arapaho National Wildlife Refuge & Illinois River Moose Viewing Site

The irrigated meadows, open ponds and river within this sanctuary are excellent habitat for Moose, muskrats, beavers and birds. Adjacent sagebrush flats and grassy knolls provide habitat for Pronghorn, Wyoming Ground Squirrels and Black-tailed Jackrabbits.

Rocky Mountain National Park

With more than 300 mi (480 km) of trails, hikers can take in many of the park's rocky peaks, clear streams, 147 lakes and plentiful flora and fauna. Interpretive programs are offered from June to October. Visitors traveling via Trail Ridge Road (U.S. 34) will see colorful alpine plants. Even though the tundra zone offers less than 40 frost-free growing days, upwards of 150 plant species may be found competing for survival in the high alpine.

Golden Gate Canyon State Park

This foothills park just west of Denver is an excellent spot for meeting summer songbirds. The visitor center has a trout viewing pond, and Ralston Creek has a beaver viewing deck. Numerous trails lead to flower-filled meadows and through mixed woodlands of conifer and vibrant deciduous aspen.

Big Blue & Powderhorn Wilderness Areas

Tall, stunning peaks, beautiful lakes, fragrant wildflowers and abundant wildlife characterize these wilderness areas. Trailheads are generally via U.S. Forest Service Trails. Lower-elevation trails are often used by 4x4 vehicles, while higher elevations are generally less disturbed and are alive with the sounds of wildlife. Elk, pikas, chipmunks, marmots and the occasional black bear may be found in these areas.

Great Sand Dunes National Monument

Night-time is the right time to visit the Great Sand Dunes to meet kangaroo rats, Mountain Cottontails, Coyotes, Bobcats and Striped Skunks. Midday views are dominated by stubborn plants growing on dune edges and by sightings of Rock Squirrels.

NEW MEXICO

Bandelier National Monument

Here, the lower, desert-like communities are full of sagebrush and cactus, leading up to pinon-juniper woodlands that finally yield to high-elevation forests dominated by Engelmann spruce and white fir. These communities, along with grasslands, shrublands and riparian canyons, provide excellent habitat for Rock Squirrels, Desert Cottontails and many other mammals.

About This Book

This guide describes 126 species of "wild" mammals that have been reported from the Rocky Mountain region. Feral populations of some domestic mammals have occurred in parts of the mountains, but, apart from the Horse (which has distinct historic and charismatic attributes), they are not described here. Humans, a member of the order Primates, have lived in the Rockies at least since the end of the last Pleistocene glaciation, but the relationship between our species and the natural world is well beyond the scope of this book, and in terms of identifying features, all you really need is a mirror.

Organization

Biologists divide mammals into a number of subgroups, called orders, which form the basis for the organization of this book. Eight mammalian orders have wild representatives in the Rocky Mountains: hoofed mammals (Artiodactyla and Perissodactyla), carnivores (Carnivora), rodents (Rodentia), hares and pikas (Lagomorpha), bats (Chiroptera), insectivores (Insectivora) and marsupials (Marsupialia). In turn, each order is subdivided into families, which group together the more closely related species. For example, within the hoofed mammals, the White-tailed Deer and the Moose, which are both in the deer family, are more closely related to each other than either is to the Pronghorn, which is in its own family.

Although the international zoological community closely monitors the use of scientific names for animals, common names, which change with time, local language and usage, are more difficult to standardize. In the case of birds, the American Ornithologists' Union has been very effective at standardizing the common names used by professionals and recreational naturalists alike. There is, as yet, no similar organization to oversee and approve the common names of mammals in North America, which can leacd to some confusion.

True moles do not occur in the Rocky Mountain region, but many people apply that name to the Northern Pocket Gopher, a burrowing mammal that leaves loose cores of dirt in fields and reminded early settlers of the moles they knew in the east and in Europe. To add to the confusion, most people use the name "gopher" to refer not to the Northern Pocket Gopher, but to the ubiquitous Richardson's Ground Squirrel. If you were to venture out of the Rockies to Minnesota, it would get even worse. There, a "gopher" is neither the Richardson's Ground Squirrel nor the Northern Pocket Gopher, but rather that state's most commonly encountered rodent, the Thirteen-lined Ground Squirrel.

You may think that such confusion is relegated to the less charismatic species of mammals, but even some of the best-known animals are victims of human inconsistencies. Most people clearly know the identity of both the Moose and the Elk, but these names can cause great confusion for European visitors. The species that we know as the Elk, *Cervus elaphus*, is called the Red Deer in Europe, where "Elk" is the named used for *Alces alces* ("elk" and *alces* come from the same root), which is known as the Moose in North America. We don't have to take the blame for this confusion, however, because early European settlers were the ones who misapplied the name "Elk" to populations of *Cervus elaphus*. In an as-yet-unsuccessful attempt to resolve the confusion, many naturalists use the name "Wapiti" for the species *Cervus elaphus* in North America. There is a small amount of hometown pride involved, as well: "Wapiti" derives from the Shawnee name for that animal, just as "Moose" is from Algonquian and "Caribou" is from Micmac.

Despite the absence of an "official" list of mammal names, there are some widely accepted standards, such as the "Revised checklist of North American mammals north of Mexico, 1997" (Jones et al. 1997, Occasional Papers, Museum of Texas Tech University, No. 173), which our book follows for both the scientific and common names of the mammals (except in regard to the common names of the *Myotis* bats).

Range Maps

Mapping the range of a species is a problematic endeavor: mammal populations are continually expanding and reducing their distributions, and dispersing individuals are occasionally encountered in unexpected areas. The range maps included in this book are intended to show the distribution of breeding/sustaining populations in the region, and not the extent of individual specimen records. Full color intensity on the map indicates a species' presence; pale areas indicate its absence.

Similar Species

Before you finalize your decision on the species identity of a mammal, check the "Similar Species" section of the account; it briefly describes other mammals that could be mistakenly identified as the species you are considering. By concentrating on the most relevant field marks, the subtle differences between species can be reduced to easily identifiable traits. As you become more experienced at identifying mammals, you might find you can immediately short-list an animal to a few possible species. By consulting this section you can quickly glean the most relevant field marks to distinguish between those species, thereby shortcutting the identification process.

Colorado Chipmunk

The MAMMALS

HOOFED MAMMALS

These mammals are the "megaherbivores" of the Rocky Mountains: they all fall into the largest size class of mammals, and they eat plants exclusively. All our native hoofed mammals belong to the order Artiodactyla (even-toed hoofed mammals). They have either two or four toes on each limb. If there are four toes, the outer two, which are called "dewclaws," are always smaller and are located higher on the leg. Horses, which belong to the order Perissodactyla (odd-toed hoofed mammals), have just a single toe on each foot. Another difference between the two orders of hoofed mammals is in the structure of their ankle bones. The ankles of even-toed hoofed mammals allow them to rise from a reclining position with their hindquarters first, which makes their large hindleg muscles more readily available for action. Horses must rise front-first. Additionally, even-toed hoofed mammals have incisors only on the lower jaw; they have a cartilaginous pad at the front of the upper jaw instead of teeth.

Cattle Family (Bovidae)

Our native bovids are distinguished by the presence of true horns in both the male and female. The horns are never shed, and they grow throughout the animal's life. They consist of a keratinous sheath (keratin is the main type of protein in our fingernails and hair) over a bony core that grows from the frontal bones of the skull. Bovids are cud chewers, and they have complex, four-chambered stomachs to digest their meals.

Bighorn Sheep

Pronghorn

Pronghorn Family (Antilocapridae)

This exclusively North American family contains just the one species. The Pronghorn has only two toes (no dewclaws), and it lacks upper canines as well as upper incisors. Unlike bovids, Pronghorns shed and regrow the keratinous sheaths of their horns each year (the bony core is not shed). Females may or may not have horns.

Deer Family (Cervidae)

All adult male cervids (and female Caribou) have antlers, which are bony outgrowths of the frontal skull bones that are shed and regrown annually. In males with an adequate diet, the antlers generally get larger each year. New antlers are soft and tender, and they are covered with "velvet," a layer of skin with short, fine hairs and a network of blood vessels to nourish the growing antlers. The antlers stop growing in late summer, and as the velvet dries up the deer rubs it off. Cervids are also distinguished by the presence of scent glands in pits just in front of the eyes. Their lower canine teeth look like incisors, so there appear to be four pairs of lower incisors.

Moose

Horse Family (Equidae)

All members of this family, which also includes zebras and donkeys, have a single toe on each limb, a bushy dorsal mane and a long, well-haired tail. Although horse-like animals were once native to North America, they disappeared from our continent more than 10,000 years ago. The herds of feral Horses that are now found in several places through the Rockies and the American West are descended from domestic Horses.

Horse

American Bison

Bos bison

Historically, millions of American Bison lived in North America. From the top of a ridge or hill, the view could be staggering—an immense herd of hundreds to thousands of bison roaming together would darken the otherwise pale landscape. Until the late 19th century, few areas of North America escaped the influence of bison: they left their impressive marks on the landscape from the northern and eastern forests, across the Great Plains and into the Rocky Mountains. Evidence of their once-great presence can still be found, even where bison no longer roam. Stained bones spill yearly from river-banks, and large boulders isolated on the plains are often smoothly polished and set in shallow pits from thousands of years of itchy bison rubbing their hides for relief.

Today, American Bison live primarily in protected areas and on private ranches. The largest herds were once found in Wood Buffalo National Park in northern Alberta, where wolves and bison continue their long-standing predator-prey relationship, but Yellowstone National Park may hold the largest wild herds today. Smaller herds can be found throughout the Rocky Mountains in areas of suitable protected habitat.

Yellowstone was the last refuge of wild bison in the U.S. at the end of the 19th century. At that time, there were fewer than 500 American Bison in all of North America. In a recovery effort, Yellowstone's native herds were built up with bison from private ranches. Continued breeding and reintroduction programs have spread the American Bison to several other areas in the Rockies, such as Grand Teton National Park, Waterton Lakes National Park and the National Bison Range. Some of the herds are fenced and heavily managed, but others are free-roaming. In Yellowstone, the bison herds are prospering, and many thousands of visitors go there each year to see these impressive animals.

The American Bison, a truly majestic and symbolic animal, has become a symbol of the difficulties involved in trying to "manage" nature. The bison in Yellowstone and Wood Buffalo are carriers of the disease brucellosis, which, if transmitted to domestic cattle, causes the cows to miscarry their young. There has never been a confirmed case of brucellosis transmission from wild bison to domestic cattle—cattle would have to come into contact with either infected birthing material or a wet newborn calf of an infected bison to contract the disease—

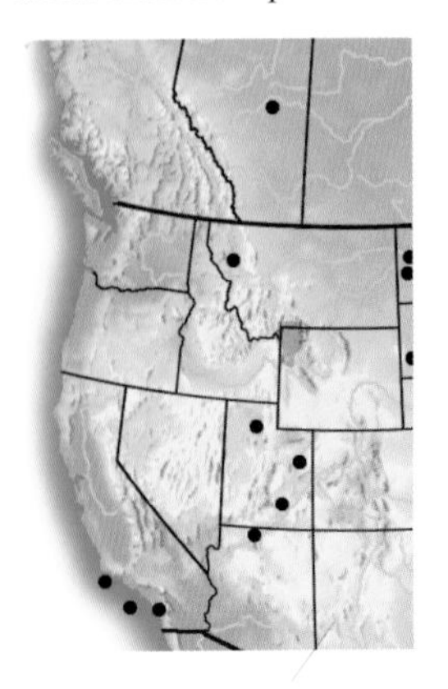

RANGE: The bison's range historically extended from the southeastern Yukon south to northern Mexico and east to the Appalachian Mountains. Free-ranging herds (shown at left) are now almost exclusively restricted to protected areas, and many small herds are raised in fenced game ranches.

DID YOU KNOW?

If bison are caught by a storm away from shelter, they face into the wind, using the woolly coat of their head and shoulders to reduce the chill.

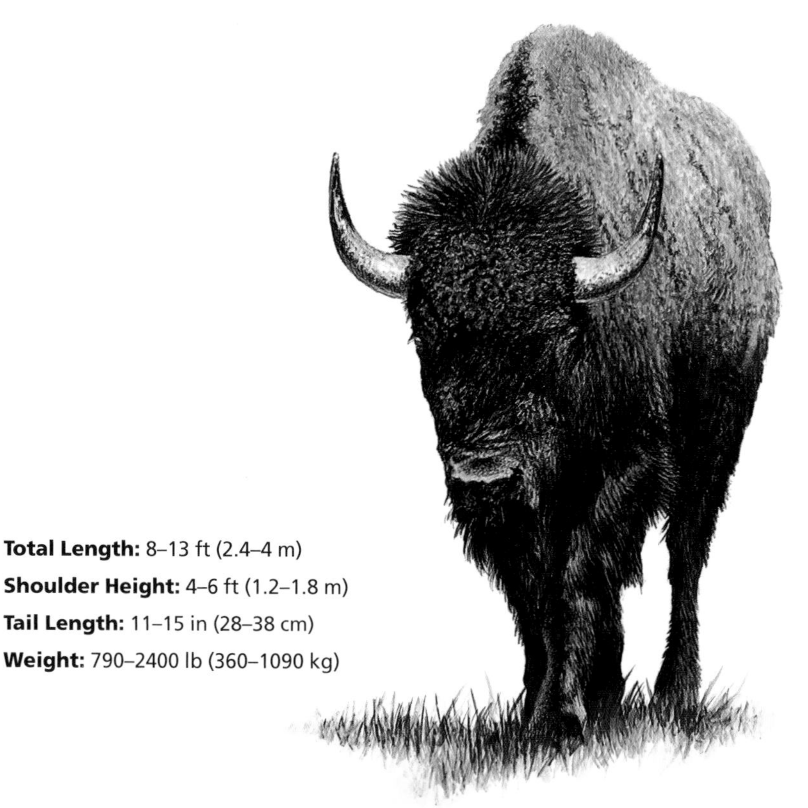

Total Length: 8–13 ft (2.4–4 m)
Shoulder Height: 4–6 ft (1.2–1.8 m)
Tail Length: 11–15 in (28–38 cm)
Weight: 790–2400 lb (360–1090 kg)

but bison, especially the Yellowstone bison, are subject to winter slaughter if they leave park boundaries. Understandably, major debates occur between the advocates of bison protection and people worried about brucellosis transmission to cattle. During the severe winter of 1997, much of the population of bison in Yellowstone moved out of the protection of the park. Several hundred of them were shot. The remainder were captured and sent for slaughter to several sites in Montana because of the brucellosis concern. It is interesting that Elk that leave the park during severe winters are still protected by game laws but bison are not.

ALSO CALLED: American Buffalo.

DESCRIPTION: The head and forequarters are covered with long, shaggy, woolly, dark brown hair that abruptly becomes shorter and lighter brown behind the shoulders. The head is massive and appears to be carried low because of the high shoulder hump and massive forequarters. Both sexes have short, round, black horns that curve upward. The legs are short and clothed in shaggy hair. The tail is long and has a tuft of hair at the tip. A bison calf is reddish at birth but becomes darker by its first fall.

HABITAT: Although the American Bison was historically most abundant in grasslands, it also inhabited alpine tundra, areas of montane and boreal forest and aspen parkland with abundant short vegetation.

FOOD: Most of the diet is made up of grasses, sedges and forbs. In winter, the

American Bison sometimes browses on shrubs, cattails and lichens, but grasses are still the primary food. A bison will paw away the snow or push it to the side with its head if the snow is not too crusted.

DEN: Historically, the American Bison was nomadic, so it did not have a permanent den. It typically beds down at night and during the hottest part of the day to ruminate. After a herd has been in an area for a while, it will leave behind wallows—dusty, saucer-like depressions where the bison rolled and rubbed repeatedly.

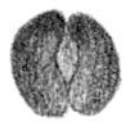

YOUNG: After a gestation of 9 to 10 months, a cow bison typically gives birth to a single, 40-lb (18-kg) calf in May. The calf is able to follow the cow within hours of birth. It begins to graze at about one week, but it is not weaned until it is about seven months old. A cow typically mates for the first time when she is two or three years old. A bull is sexually mature then, too, but competition from older males customarily prevents him from breeding until he is seven to eight years old.

walking trail

SIMILAR SPECIES: No other native Rocky Mountain mammal resembles an American Bison. The Moose (p. 60) has a similarly colored coat, but it is taller and has long, thin, light-colored lower legs and a much longer and leaner body overall. A bull Moose has broad antlers, not horns.

Moose

Mountain Goat

Oreamnos americanus

Acrophobia—the fear of heights—is a mystery to the Mountain Goat. The Mountain Goat is the Rocky Mountains' foremost natural mountaineer, and the very heights that instill fear in so many people are comfortable and easily navigable for this animal. As a consequence of its habitat preference, the Mountain Goat has the added bonus of living its entire life surrounded by the sweeping scenery of the Rocky Mountains.

One of the most famous places in the Rockies to see Mountain Goats is the Going to the Sun Highway in Glacier National Park. There you will not only experience goats in their natural habitat, you will see them against the backdrop of some of the most awe-inspiring peaks and valleys of the Rockies. Another excellent goat viewing site is the Icefields Parkway between Banff and Jasper. Introduced populations occur in the Yellowstone area, but they are only occasionally seen by visitors.

The Mountain Goat has several physical characteristics that help it live in such precarious situations. The hard outer ring of its hooves surrounds a softer, spongy, central area that provides a good grip on rocky surfaces. The dewclaws are long enough to touch the ground on soft surfaces, and they provide greater "flotation" on weaker snow crusts. To keep it relatively comfortable in the subzero temperatures and strong winter winds that sweep along mountain faces, the Mountain Goat's winter coat consists of a thick, fleecy undercoat surmounted by guard hairs that are more than 6 in (15 cm) long. By the time the warmth of June arrives, goats begin to shed "blankets" of thick hair, often in their dusting pits dug high on the sides of mountains. The fur falls off in pieces, and during early summer, when many tourists visit the mountain parks, Mountain Goats are not in their picturesque prime. Their short, neat, white summer coat comes in by July, and it continues to grow to form the thick winter coat.

The steep relief of its rocky home offers many protective qualities for a Mountain Goat, but the ever-present risk of avalanches is an expensive trade-off. Snowslides are a major cause of death among most populations of Mountain Goats, particularly during late winter and spring. These tragic incidents are not universally cursed, however, because recently awakened, winter-starved bears and hungry Wolverines forage along spring slides for the snow's victims.

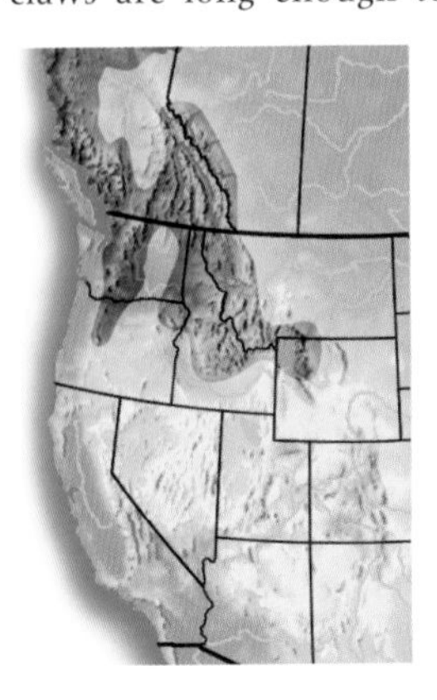

RANGE: This goat's natural range extends from southern Alaska and the eastern Yukon south through the Coast Mountains into the Washington Cascades and southeast through the Rockies into Idaho and Montana. It has been successfully introduced to several locations in the western states.

DID YOU KNOW?

The Mountain Goat's skeleton is arranged so that all four hooves can fit on a ledge as small as 6 in (15 cm) long and 2 in (5 cm) wide. A goat can even rear up and turn around on such a tiny foothold.

Total Length: 4–5 ft (1.2–1.5 m)

Shoulder Height: 3–4 ft (0.9–1.2 m)

Tail Length: 3½–5½ in (8.9–14 cm)

Weight: 100–300 lb (45–136 kg)

DESCRIPTION: The coat of this stocky, hump-shouldered animal is white and usually shaggy, with a longer series of guard hairs overtop a fleecy undercoat. The lips, nose, eyes and hooves are black. Both sexes may sport a noticeable "beard," which is longer in winter. The short legs often look like they are clothed in breeches in winter, because the hair of the lower leg is much shorter than that of the upper leg. The tail is short. The ears are relatively long. Both sexes have narrow, black horns. A billy's horns are thicker and curve backward along a constant arc. A nanny's horns are narrower and tend to rise straight from the skull and then bend sharply to the rear near their tips. A Mountain Goat kid is also white, with a gray-brown stripe along its back.

HABITAT: The Mountain Goat generally occupies steep slopes and rocky cliffs in alpine or subalpine areas, where low temperatures and deep snow are common. Although it typically inhabits treeless areas, the Mountain Goat may travel through dense subalpine or montane forests going to and from salt licks. In summer, it tends to be seen more frequently at lower elevations, especially in flower-filled alpine meadows not far from the escape shelter of cliffs. It moves to the highest windswept ledges in winter to find vegetation that is free of snowcover.

FOOD: This adaptable herbivore varies its diet according to its environment: in a Montana study, it ate shrubs almost exclusively, with the balance of the diet coming from mosses, lichens and forbs; in Alberta, only one-quarter of the diet was shrubs and three-quarters was grasses, sedges and rushes. The Mountain Goat's winter feeding areas are generally separate from the summer areas. At about the same time as its early summer molt, the Mountain Goat has a strong need for salt, and it may travel long distances to find outcrops of mineral-rich soil.

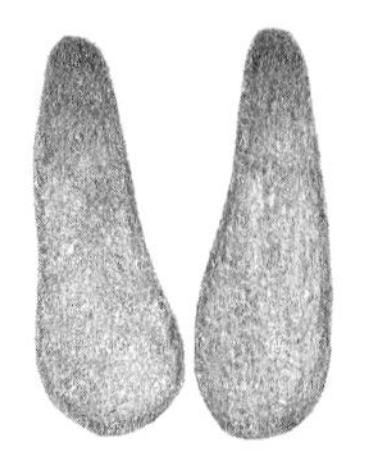

hoofprint

DEN: Mountain Goats bed down in shallow depressions scraped out in shale or dirt at the base of a cliff. Clumps of the goats' white hair are often scattered in the vicinity of the scrapes. In early summer, goats dig dusting pits where they rub and sit in state like big dogs. Nannies will often evict billies from their dusting sites, revealing their dominant status.

YOUNG: In May, after a gestation of five to six months, a nanny bears a single kid (75 percent of the time) or twins, weighing 6½–8½ lb (2.9–3.9 kg). A kid can follow its mother within hours. After a few days, the kid starts eating grass and forbs, but it is not weaned until it is about six weeks old. The young are very playful, and they leap, jump and eagerly scale boulders as they learn their art of rock climbing. Both sexes become sexually mature after about 2½ years. A nanny will mate every other year.

walking trail

SIMILAR SPECIES: The Bighorn Sheep (p. 36) has brown upperparts and a whitish rump patch. Its brown horns are either massive and thick at the base (in rams) or flattened (in ewes), but never round, thin, stiletto-like or black like a Mountain Goat's horns.

Bighorn Sheep

Bighorn Sheep

Ovis canadensis

No matter where you travel in North America, the Rockies simply cannot be beat for their diversity of hoofed mammals. It seems fitting, therefore, that one of the most recognizable and revered ungulates, the Bighorn Sheep, is a favorite symbol of the Rocky Mountain wilderness. Although Bighorn Sheep can routinely be seen along roadsides in parks and preserves, they have a well-developed sense of balance, and it is along steep slopes and rocky ledges that they seem to belong.

Now that the days of hunting Bighorn Sheep in protected areas have long passed, many animals wander comfortably around areas of human activity. Provided that people are unobtrusive and non-aggressive, they can be rewarded with glimpses of the sheep's natural behavior amidst the beautiful mountain scenery. Some of the best places to experience the majesty of Bighorn Sheep include Mammoth Hot Springs in Yellowstone National Park, Going to the Sun Highway in Glacier National Park and roadside cliffs in Jasper and Banff national parks. As friendly and quiet as a Bighorn Sheep appears, however, always remember that it is a wild animal and deserves to be treated as such.

Bighorn lambs that are too young and too small to have mastered the sanctuary of cliffs are particularly vulnerable to Coyotes and Gray Wolves. Newborn lambs occasionally become prey for eagles, Mountain Lions and Bobcats as well. Provided they survive their first year, however, most Bighorns live long lives—few of their natural predators have the ability to match the Bighorn Sheep's sure-footedness and vertical agility.

The magnificent courtship battles between Bighorn rams have made these animals favorites of TV wildlife specials and corporate advertising. During October and November, adult rams establish a breeding hierarchy that is based on the relative sizes of their horns and the outcomes of their impressive head-to-head combats. In combat, opposing rams rise on their hindlegs, run a few steps toward one another and smash their horns together with glorious fervor. Once the breeding hierarchy has been established, mating takes place, after which the rams and ewes tend to split into separate herds. For the most part, the rams abandon their head blows until the next fall, but broken horns and ribs are reminders of their hormone-induced clashes.

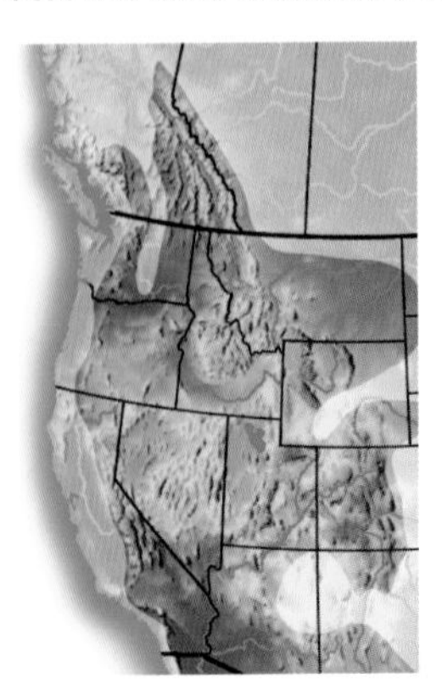

RANGE: From the Rocky Mountains of Alberta and the Coast Mountains of southwestern British Columbia, the Bighorn Sheep's range extends east through Montana and south through California and New Mexico into northern Mexico.

DID YOU KNOW?

Bighorn rams occasionally interbreed with domestic ewes. A number of such crosses have occurred in the southern Rockies, where domestic sheep flocks range into the mountains. The hybrids have the economically inferior, coarse hair of Bighorns.

Total Length: 5–6 ft (1.5–1.8 m)

Shoulder Height: 30–45 in (76–114 cm)

Tail Length: 3¼–5 in (8.3–13 cm)

Weight: 120–340 lb (54–154 kg)

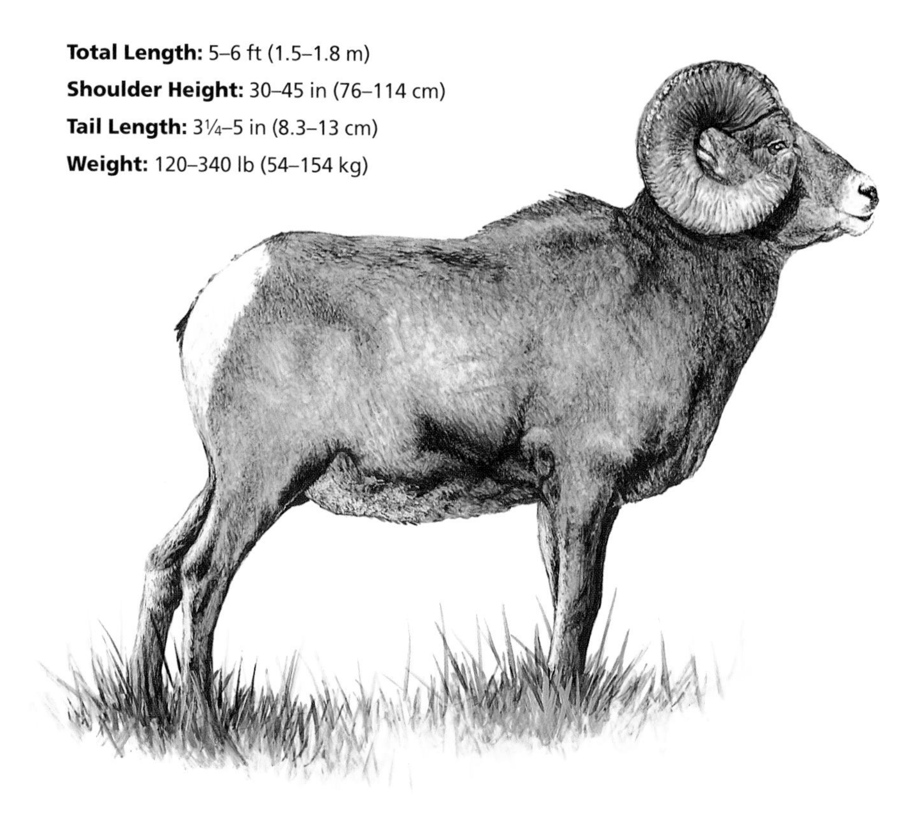

ALSO CALLED: Mountain Sheep.

DESCRIPTION: This robust, brownish sheep has a bobbed tail and a large, white rump patch. The belly, the insides of the legs and the end of the muzzle are also white. The brown coat is darkest in fall, gradually fading with winter wear. It looks motley in June and July while the new coat grows in. "Bighorn" is a well-deserved name, because the circumference of a ram's horns can be as much as 18 in (46 cm) at the base. The curled horns can be 43 in (109 cm) long and spread 26 in (66 cm) from tip to tip. Heavy ridges, the pattern of which is unique to each individual, run transversely across the horn. A deep groove forms each winter, which makes it possible to determine a sheep's age from its horns. A ewe's horns are shorter and noticeably more flattened from side to side than a ram's. Also, a ewe's horns never curl around to form even a half circle, whereas an older ram's horns sometimes form a full curl or more.

HABITAT: Although it is most common in non-forested, mountainous areas, where cliffs provide easy escape routes, the Bighorn Sheep can thrive outside the mountains as long as precipitous slopes are present in the vicinity of appropriate food and water. Some populations live along steep riverbanks and even in the gullied badlands of desert environments.

FOOD: The diet consists primarily of broad-leaved, non-woody plants and grasses. Exposed, dry grasses on windswept slopes provide much of the winter food. The Bighorn Sheep exhibits a ruminant's appetite for salt—to fulfill this need, herds may travel miles, even through dense forests, to reach natural salt licks. They often eat soil alongside highways for the road salt that is applied during winter. This activity unfortunately increases the number of collisions with vehicles.

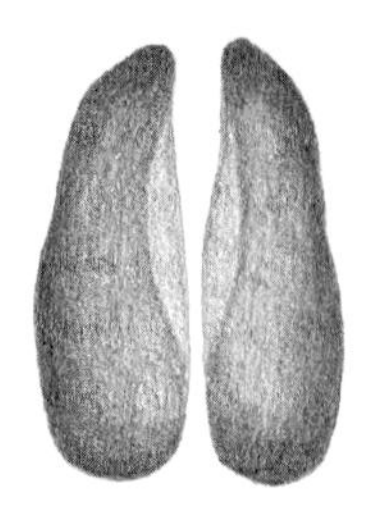

hoofprint

DEN: A Bighorn Sheep typically beds down for the night in a depression that is about 4 ft (1.2 m) wide and up to 1 ft (30 cm) deep. The depression usually smells of urine and is almost always edged with the sheep's tiny droppings.

walking trail

YOUNG: Typically, a ewe gives birth to a single lamb in seclusion on a remote rocky ledge in late May or early June, after a gestation period of about six months. The ewe and her lamb rejoin the herd within a few days. Initially, the lamb nurses every half hour; as it matures, it nurses less frequently, until it is weaned at about six months old. Lambs are extremely agile and playful: they jump and run about, scale small cliffs, engage in mock fights and even jump completely over one another. These activities prepare them for escaping predators later in life.

SIMILAR SPECIES:
The Dall's Sheep (p. 40) has thinner horns and occurs to the north of the Bighorn's range. The Mule Deer (p. 52) also has a large, whitish rump patch and an overall brown color, but bucks typically have branched antlers, and does have no head protrusions (other than their ears). The Mountain Goat (p. 32), which sometimes shares habitat with the Bighorn Sheep, is white, not brown, and its horns are black and cylindrical.

Dall's Sheep

Dall's Sheep

Ovis dalli

In the mountains and highlands of the northernmost Rockies, Dall's Sheep appear as tiny, dirty spots on a northern wilderness palette. Closely related to Bighorn Sheep, Dall's Sheep are easy to identify because of their long, wide-spreading, spiraled horns. They are sometimes called "thinhorn" sheep, because their horns are relatively thin at the base when compared to the massive horns of Bighorn Sheep. Dall's ewes also have horns, but they are much reduced in size and shape.

There are two distinct races of the Dall's Sheep, and some texts have even considered them as separate species. The northerly race has a predominantly white coat, while the more southern race of British Columbia, sometimes called the Stone Sheep, has a darker gray coat. There are areas where the two races intergrade, producing offspring with gray backs and white heads, legs and rumps.

As the seasons change, the composition of a herd of Dall's Sheep changes as well. Ewes and lambs form nursery groups from early summer until fall. During this time, the rams journey high into the mountains. Their separation from the ewes and lambs means that each group has less competition for food. The ewes remain in the better grazing areas, so they can best nourish both themselves and their young. Higher in the mountains, some of the rams may group together, but often the oldest and dominant rams remain solitary.

In fall, the two groups come together for the mating season, which is a busy and dramatic time for Dall's Sheep. The rams engage in vigorous courtship battles to determine their status. Competing rams rise up on their hindlegs and lunge forward into their opponent in the same manner as Bighorn Sheep. With their heads lowered, the rams crash their horns together. After a few head-on blows, the rams push and shove each other until one of them turns away. The dominant ram wins the chance to mate with the most ewes.

Bands of Dall's Sheep must often cross extensive lowlands as they travel from their summer ranges to their winter ranges. During this time they are away from the safety of the cliffs, and in open, gentle terrain they are vulnerable to predation by Gray Wolves, Mountain Lions, Canada Lynx, Wolverines and bears. On occasion, a Golden Eagle may swoop down and take a young lamb.

ALSO CALLED: Thinhorn Sheep.

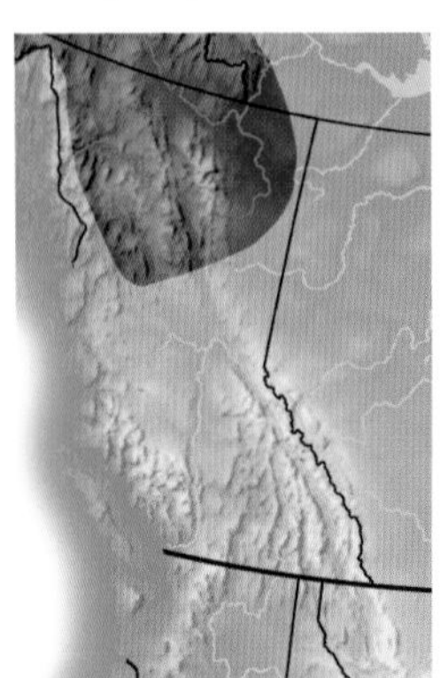

RANGE: Dall's Sheep occur in all but the extreme northern and western parts of Alaska, across the Yukon, in the western mountains of the Northwest Territories and in the northern mountains of British Columbia.

DID YOU KNOW?

When Dall's Sheep rams engage in their fall head-butting contests, the sound of their horns clashing together can be heard more than a mile away.

Total Length: 4½–6 ft (1.4–1.8 m)
Shoulder Height: 30–41 in (76–104 cm)
Tail Length: 2¾–4½ in (7–11 cm)
Weight: 100–220 lb (45–100 kg)

DESCRIPTION: The Dall's Sheep that are found in the Rockies of north-central British Columbia are slate brown to almost blackish overall, except for the white on the muzzle, forehead, rump patch and inside of the hindlegs. The horns and hooves are light amber. The iris of the eye is golden brown. A ram has thicker horns than a ewe, and they spiral widely. The horns of a ewe are short and curved backward, never achieving the complete spirals sometimes exhibited by a ram's horns.

HABITAT: In summer, Dall's Sheep occupy alpine tundra slopes to an elevation of 6600 ft (2000 m). They descend to drier south- or southwestern-facing slopes in winter. Bands of Dall's Sheep may travel long distances outside their typical habitats to find mineral licks.

DEN: The Dall's Sheep does not keep a den, but it is seldom far from steep, rocky cliffs, which serve as escape cover from eagles and carnivores. At night, a Dall's Sheep beds down wherever it is, choosing an elevated site with good visibility. In rocky areas, it will paw the ground to remove the larger stones and create a gravelly bed. Sometimes it will bed down in a meadow or on a roughened site formed where a Grizzly Bear dug for food.

FOOD: Broadleaf herbs are favored in spring and summer, with grasses and seeds making up most of the winter diet. The tips of willow, pasture sage, cranberry, crowberry and mountain avens are also consumed in winter.

YOUNG: Usually a single lamb, occasionally twins, is born in the second or third week of May, following a gestation of slightly less than six months. The lambs lie close to their mother at first, but within a few days they are climbing about the cliffs. By the time they are a month old, the lambs form groups and begin to feed on plants, but they continue to nurse for nine months. A ewe first breeds in her second fall, and she may mate with several rams during the day or two when she is receptive. A ram is typically seven to eight years old before he gets a chance to mate.

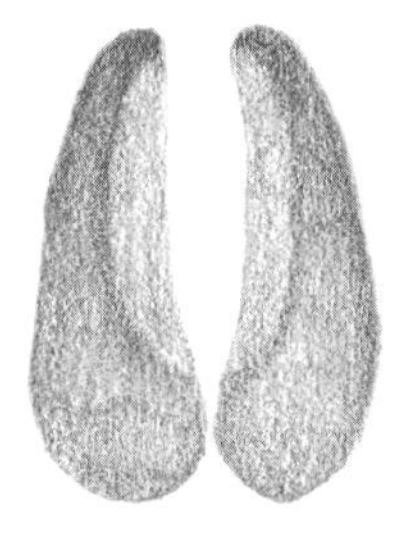

hoofprint

walking trail

SIMILAR SPECIES: The Bighorn Sheep (p. 36) is brownish overall, it occurs to the south of the Dall's Sheep's range, and the ram has more massive horns. The Mountain Goat (p. 32) is all white and has black, stiletto-like horns, longer fur and often a "beard."

Bighorn Sheep

Pronghorn

Antilocapra americana

Occasionally, on a dry, grassy slope in the mountains or foothills, the shape of a Pronghorn will emerge from the fawn-colored landscape to stand and stare. Just as suddenly, it turns and retreats, blending into the grasses that wave on the hillside.

Pronghorns, which once roamed the North American plains 40 million strong, are found in smaller numbers in the Rocky Mountains than on the Great Plains. In the Yellowstone area alone, however, there are at least 5000 of these striking animals. Some of the best places to view Pronghorns in the Rockies are the Mammoth Hot Springs area of Yellowstone, the Seedskadee National Wildlife Refuge in Wyoming and Arapaho National Wildlife Refuge in Colorado.

The Pronghorn superficially resembles a deer, and it is often mistakenly called an antelope, but it shares no family ties to either—it is the sole surviving species of a family of North American hoofed mammals whose fossil record dates back 20 million years. This animal's unique, pronged horns are not antlers, and only the outer keratin sheath, not the bony core, is shed each year.

Should danger be perceived, a Pronghorn will erect the hairs of its white rump patch to produce a mirror-like flash that is visible at great distances. Speed, which comes easily and quickly to the Pronghorn, is this animal's chief defense. The Pronghorn is the swiftest of North America's land mammals, and among the fastest in the world. With its efficient metabolism, powered by an extremely large heart and lungs for its body size, the Pronghorn can run at highway speeds for several minutes at a time. Its lack of dewclaws is thought to be an adaptation for speed.

Young Pronghorns are born all legs and ready to go. Within just a few minutes after birth a fawn is up and walking, although it spends most of its time lying motionless. In five days, it can outrun an Olympic sprinter, and when it is three weeks old it can keep up with its mother, and they rejoin the herd. For those first few days before the fawn is really fast, however, its survival trick is its smell—or the lack of it—and its ability to lie flat and motionless. A Coyote passing upwind may not notice a fawn lying just a few feet away.

For all its speed, the Pronghorn is a poor jumper. Its numbers declined rapidly with the settling and fencing of open land throughout the West and the hunting of hoofed mammals for food. In the

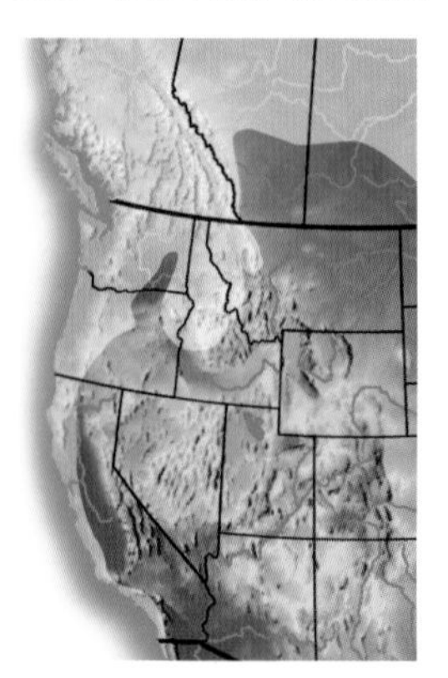

RANGE: The Pronghorn is found through much of western North America, from southern Alberta and Saskatchewan southwest into Oregon and south through California and western Texas into northern Mexico.

DID YOU KNOW?

The eyes of a Pronghorn are set so far to the sides of its head that it can see behind itself as well as in front. This wide field of vision is undoubtedly useful in detecting and avoiding predators.

Total Length: 4–4½ ft (1.2–1.4 m)
Shoulder Height: 32–41 in (81–104 cm)
Tail Length: 3⅜–6 in (8.6–15 cm)
Weight: 70–140 lb (32–64 kg)

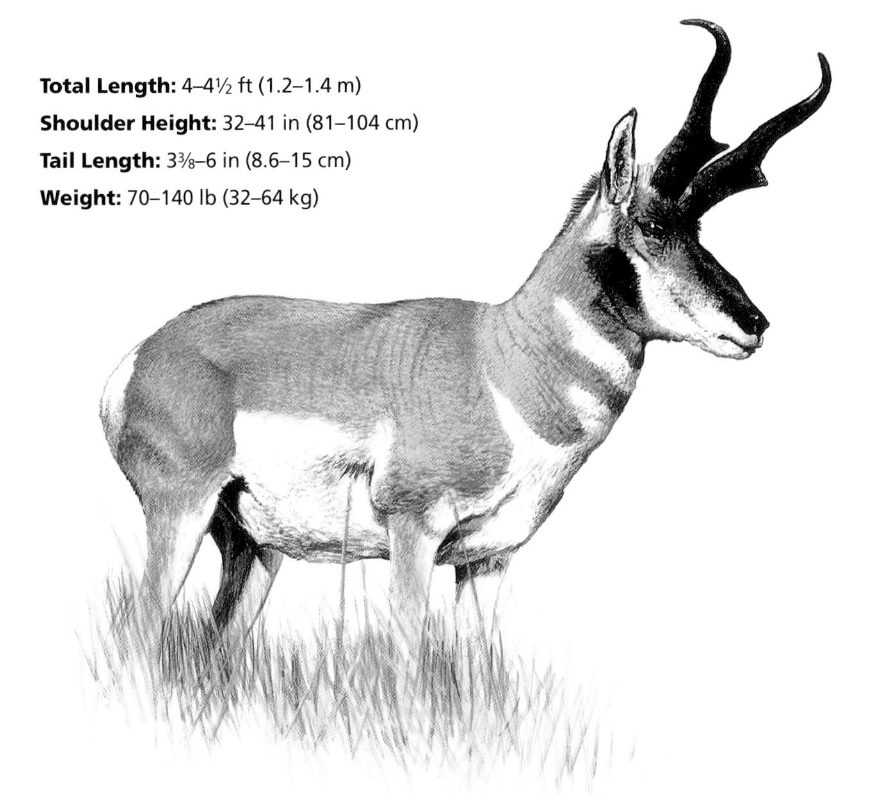

mid-1900s, several reserves were established to protect dwindling Pronghorn populations. Fences still remain throughout much of the Pronghorn's range, but many fences are now constructed to leave enough room for the animals to fit underneath. Running Pronghorns surprise many a passing motorist when, one after another, they hardly break stride to deftly dip beneath the lowest strand of barbed wire in a fence. Sometimes, a Pronghorn's loosely attached hair comes off in a cloud when it hits a wire.

ALSO CALLED: American Antelope.

DESCRIPTION: The upperparts, legs and tail are generally tan. The belly, lower sides and lower jaw are white, and there are two broad white bands across the throat and a large white rump patch. Both sexes may have horns, but those of a doe are never as long as her ears, and they do not have the ivory-colored tips seen on a buck. A buck's horns are straight near the base and then bear a short branch, or "prong," before they usually curve backward or inward to sharp tips. The muzzle is black, and on a buck the black extends up over the face to the horn bases. A buck also has a broad black stripe running from the ear base to behind the lower jaw. There is a short, black-tipped mane on the nape of the neck. There are no dewclaws on the legs.

HABITAT: The Pronghorn is a staunch resident of treeless areas. It inhabits open, often arid grasslands, grassy brushlands and semi-deserts along the edge of the mountains. It avoids forests, but grazes at high elevations in Yellowstone National Park and neighboring areas.

FOOD: The winter diet is composed almost exclusively of sagebrush and other woody shrubs. In one study, the spring diet switched to snowbrush, snowberry, rabbitbrush and sagebrush for 67 percent of the intake, forbs for 17 percent, alfalfa and crops for about 15 percent and grasses for only 1 percent.

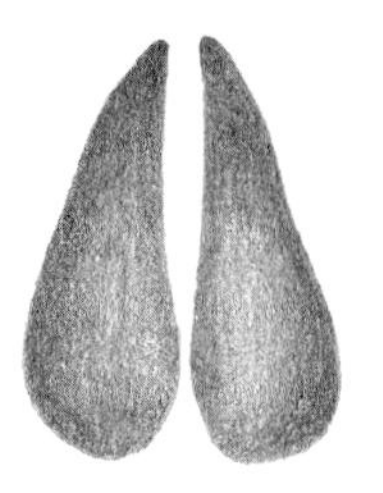
hoofprint

DEN: Because it is a roaming animal that remains active day and night—it alternates short naps with watchful feeding—the Pronghorn does not maintain a home bed.

YOUNG: Forty percent of does bear a single fawn with their first pregnancies, but 60 percent of first pregnancies and nearly all subsequent pregnancies result in the birth of twins. A doe finds a secluded spot on the prairie to give birth in June, following a gestation period of 7½ to 8 months. The fawns lie hidden in the grass at first, and their mothers return to nurse them about every 1½ hours. The does gradually reduce the frequency of nursings, and when a fawn is about two weeks old and capable of outrunning most potential predators, mother and young rejoin the herd. Some does may breed during the short, mid- to late-September breeding season of their first year, but most do not breed until their second year.

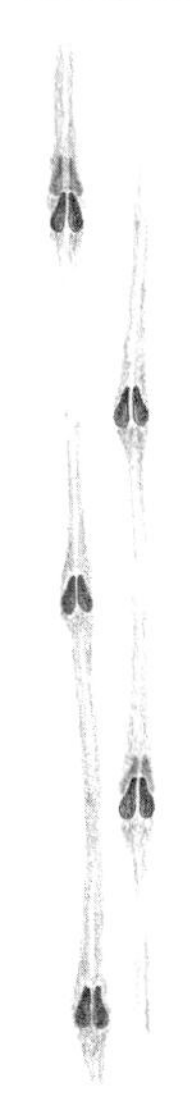
walking trail

SIMILAR SPECIES: The Mule Deer (p. 52) also has a white rump, but it is larger, and the bucks have antlers, not black horns. The White-tailed Deer (p. 56) does not have a white rump. Neither deer has the white throat bands or white lower sides of the Pronghorn.

White-tailed Deer

Elk

Cervus elaphus

The pitched bugle of the bull Elk is, through much of the Rocky Mountains, as much a symbol of fall as the first frost, the golden aspen leaves and the honk of migrating geese. The Elk has likely always held some form of fascination for humans, as evidenced by native hunting and lore, but it is another of North America's large mammals that suffered widespread extirpation during the time of Euro-american settlement and agricultural expansion across the continent.

The dramatic decline of Elk in North America during the 19th century prompted such measures as the Canadian government's creation of Elk Island National Park in 1907. Even the great numbers of Elk currently seen in mountain parks owe their presence to mitigative human efforts. Northern populations were so depleted that reintroductions of Elk were necessary to form new herds. Animals from Yellowstone National Park in Wyoming were used to develop herds in Banff National Park in Alberta from 1917 to 1920. The large herds in the Grand Teton area are also the result of human intervention. Decades ago, concerned by a lack of winter forage, wildlife officials began to supplement the Teton Elk's winter diet. This action not only saved the local population, but it encouraged immigration and population growth to an artificially high level.

Elk form breeding harems to a greater degree than most other deer. A bull Elk that is a harem master expends a considerable amount of energy during the fall rut—his fierce battles with rival bulls and the upkeep of cows in his harem demand more work than time permits—and, if snows come early, he starts winter in a weakened state. Once the rut is over, however, bulls fatten up by as much as a pound a day. Cows and young Elk, on the other hand, usually see the first frost while they are fat and healthy. This disparity is ecologically sound: many cows enter winter pregnant with the future of the Elk population, whereas, once winter arrives, the older bulls' major contributions are past.

Fortunately for Elk, much of the Rocky Mountains has become more accessible to grazing, even during winter. Artificially lush golf courses and agricultural fields supply high-quality forage throughout the year, while roads, townsites and human activity have eliminated most major predators. Today Elk are common throughout montane regions of the Rocky Mountains, and

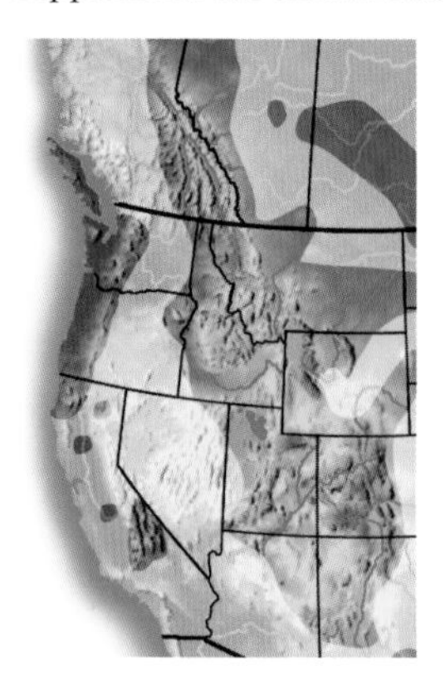

RANGE: In North America, the elk occurs from northeastern British Columbia southeast to southern Manitoba, south to southern Arizona and New Mexico and along the Pacific Coast from Vancouver Island to northern California. It has been introduced as a game species and as ranch livestock in many areas.

DID YOU KNOW?

By the end of the 1800s Elk had disappeared from eastern North America. From a low estimate of perhaps 41,000 for the entire continent, the species has since recovered to nearly 1 million. Elk are being reintroduced to some areas in the East.

Total Length: 6½–8½ ft (2–2.6 m)
Shoulder Height: 4–5 ft (1.2–1.5 m)
Tail Length: 4¾–7 in (12–18 cm)
Weight: 400–1100 lb (180–500 kg)

they are an expected sight throughout the year in Yellowstone, Banff and Jasper national parks and even as far south as New Mexico. In wilder areas, Elk are typically most active during the daytime, particularly near dawn and dusk, but they often become nocturnal in areas of high human activity where hunting occurs.

ALSO CALLED: Wapiti.

DESCRIPTION: The summer coat is generally golden brown. The winter coat is longer and grayish brown. Year-round, the head, neck and legs are darker brown, and there is a large yellowish to orangish rump patch that is bordered by black or dark brown fur. The oval metatarsal glands on the outside of the hocks are outlined by stiff yellowish hairs. A bull Elk has a dark brown throat mane, and he starts growing antlers in his second year. By his fourth year, the bull's antlers typically bear six "points" to a side, but there is considerable variation both in the number of points a bull will have and the age when he acquires the full complement of six. A bull will rarely have seven or eight points. The antlers are usually shed in March. New ones begin to grow in late April, becoming mature in August.

HABITAT: Although the Elk prefers upland forests and prairies, it sometimes ranges into alpine tundra, coniferous forests or brushlands. In the Rockies, the Elk tends to move to higher elevations in spring and lower elevations in fall.

FOOD: Elk are some of the most adaptable browsers or grazers. Woody plants and fallen leaves frequently form much of their winter and fall diet. Sedges and grasses frequently make up 80 to 90 percent of the diet in spring and summer. Salt is a necessary dietary component for all animals that chew their cud, and Elk may travel vast distances to devour salt-rich soil.

DEN: The Elk does not keep a permanent den, but it often leaves flattened areas of grass or snow where it has bedded down during the day.

YOUNG: A cow Elk gives birth to a single calf between late May and early June, following an 8½-month gestation. The young stand and nurse within an hour, and within two to four weeks the cow and calf rejoin the herd. The calf is weaned in the fall.

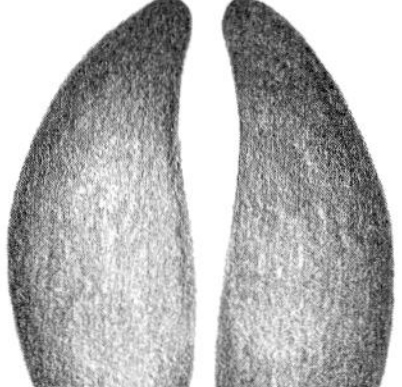

hoofprint (walking)

walking trail

SIMILAR SPECIES: The Moose (p. 60) is darker and taller and has lighter-colored lower hindlegs. The Bighorn Sheep (p. 36) and Mule Deer (p. 52) have whitish, rather than yellowish, rump patches and are generally smaller. Also, the Bighorn Sheep has curled horns, not antlers.

Moose

Mule Deer

Odocoileus hemionus

The Mule Deer has been around since prehistoric times, and it continues to thrive in the mountains and in broken and fragmented landscapes. If you want an intimate encounter with a large deer in the Rocky Mountains, there may be no better candidate than the Mule Deer. It tends to frequent open areas in parks and other protected areas, and it can be very bold, conspicuous and quite approachable. Campgrounds in the mountain parks often afford tremendous opportunities to begin collecting wildlife memories with this species.

One of the Mule Deer's best-known characteristics is its bouncing gait, which is known as "stotting" or "pronking." When it stots, a Mule Deer bounds and lands with all four legs simultaneously, so that it looks like it's using a pogo-stick. This fascinating gait allows the deer to move safely and rapidly across and over the many obstructions it encounters in the complex brush and hillsides areas it typically inhabits. Although stotting is characteristic of the Mule Deer, this animal also walks, trots and gallops perfectly well. When disturbed, a retreating Mule Deer will often stop for a last look at whatever disturbed it before it disappears completely from view.

Mule deer feed at dawn, at dusk and well into the night. They have great difficulty traveling through snow that is more than knee deep, so they are unable to occupy many high areas within the Rocky Mountains in winter. To avoid the snow, they migrate to lower elevations at the onset of winter, often into townsites, which have buried grasses and dormant ornamentals that are much to their liking.

During the mating season, Mule Deer bucks compete for the does that are in estrus. Two bucks will tangle with their antlers, trying to force the other's head lower than theirs. The weaker of the two eventually surrenders and usually leaves the area. Rarely, the antlers of two bucks become locked during these competitions, and if they are unable to free themselves, both bucks inevitably perish from either starvation, wounds that were inflict during the battle or being found by a predator.

In regions where the Mule Deer and the White-tailed Deer both occur, they do hybridize on occasion. Hybrid male offspring are sterile, and although hybrid females are fertile, all hybrids seem to have higher mortality rates than the pure species, which may be why hybrids are rarely seen.

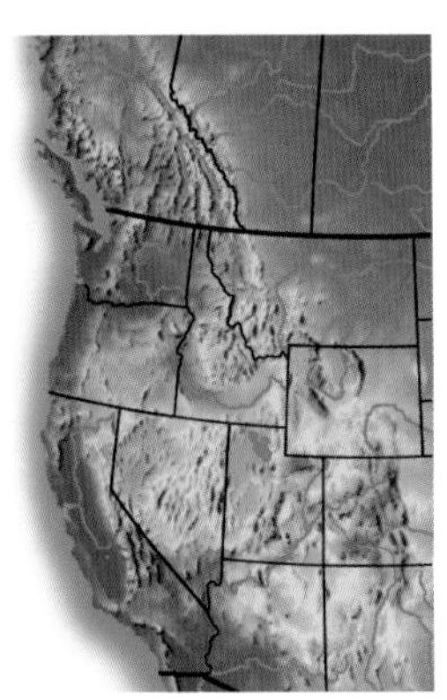

RANGE: Widely distributed through western North America, this deer ranges from the southern Yukon southeast to Minnesota and south through California and western Texas into northern Mexico.

DID YOU KNOW?

Although the Mule Deer is usually silent, it can snort, grunt, cough, roar and whistle. A fawn will sometimes bleat. Even people who have observed deer extensively may be surprised to encounter one that is vocalizing.

Total Length: 4½–5½ ft (1.4–1.7 m)

Shoulder Height: 35–41 in (89–104 cm)

Tail Length: 4¾–8¾ in (12–22 cm)

Weight: 68–470 lb (31–210 kg)

ALSO CALLED: Black-tailed Deer.

DESCRIPTION: The Mule Deer gets its name from its large, mule-like ears. It has a large, whitish rump patch that is divided by the short, black-tipped tail. The dark forehead contrasts with both the face and upperparts, which are tan in summer and dark gray in winter. There is a dark spot on either side of the nose. The throat and insides of the legs are white year-round. A buck has fairly heavy, up-swept antlers that are equally branched into forked tines. The metatarsal glands on the outside of the lower hindlegs are 4–6 in (10–15 cm) long.

HABITAT: This deer's summer habitats vary from dry brushlands to alpine tundra. Bucks tend to move to the tundra edge at higher elevations, where they form small bands; does and fawns remain at lower elevations. In drier regions, both sexes are often found in streamside situations. The Mule Deer thrives in the early successional stages of forests, so it is often found where fire or logging removed the canopy a few years before.

FOOD: Grasses and forbs form most of the summer diet. In fall, the Mule Deer consumes both the foliage and twigs of shrubs. The winter diet makes increasing use of twigs and woody vegetation, and grazing occurs in hayfields adjacent to cover.

DEN: The Mule Deer leaves oval depressions in grass, moss, leaves or snow where it lays down to rest or chew its cud. It typically urinates upon rising. A doe usually steps to one side first, but a buck will urinate in the middle of the bed.

YOUNG: Following a gestation period of 6½ to 7 months, a doe gives birth to one to three (usually two) fawns in May or June. The birth weight is 7¾–8½ lb (3.5–3.9 kg). A fawn is born with light dorsal spots, which it carries until the fall molt in August. The fawn is weaned when it is four to five months old. It becomes sexually mature at 1½ years.

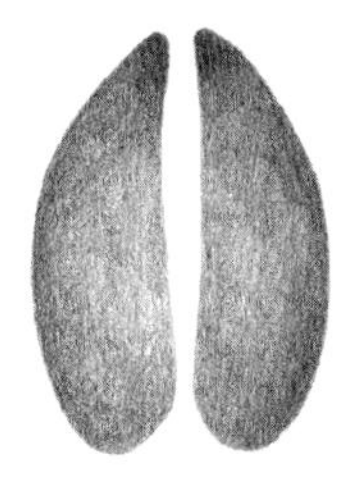

hoofprint (walking)

stotting group

SIMILAR SPECIES: The White-tailed Deer (p. 56) has a much smaller rump patch, which is usually hidden by the reddish- to grayish-brown upper surface of the tail, and much shorter metatarsal glands. It shows the white undersurface of its tail when it runs. A White-tail buck's antlers consist of a main beam with typically unbranched, rather than equally forked, tines. The Elk (p. 48) is larger, has a dark mane on the throat and has a yellowish or orangish rump patch. The Pronghorn (p. 44) has black horns and white throat bands, and the lower half of its sides are white.

White-tailed Deer

White-tailed Deer

Odocoileus virginianus

Given the current status of the White-tailed Deer in the Rocky Mountains, it is hard to imagine that before the arrival of Europeans these graceful animals were only found in small, isolated populations. Historically, these deer were rather uncommon in the Rocky Mountains, but with the spread of agricultural development and forest fragmentation, the White-tailed Deer has become much more widespread. In some parts of the Rockies, White-tailed Deer are now more commonly seen than Mule Deer.

The White-tailed Deer is a master at avoiding detection, so it can be frustratingly difficult to observe. It is very secretive during daylight hours, when it tends to remain concealed in thick shrubs or forest patches. Once the sun begins to set, however, the White-tailed Deer leaves its daytime resting spot to travel to a foraging site. The White-tailed Deer moves gracefully, weaving an intricate path through dense shrubs and over fallen trees. Indeed, a White-tailed Deer in prime form seems uncatchable in its own habitat. The animal itself clearly does not share this view—its nose and ears continually twitch, aware that any shadow could conceal a predator. Wolves, Mountain Lions and humans are the major threats to this deer, although fawns and sick individuals may be easy prey for Coyotes, too.

Speed and agility are effective against most of the White-tail's predators, but all deer are vulnerable to severe winters. Deep snow and a scarcity of high-energy food leave the deer with a negative energy budget from the time of the first deep snowfalls until green vegetation reemerges in spring. In spite of their slowed metabolic rates during winter, many deer may starve before spring arrives; in doing so, they provide food for scavengers.

In the national parks, White-tailed Deer may become habituated to the presence of humans, and they can sometimes be closely approached. Doing so can be perilous, however, especially to children, because White-tail does can rear up and strike downward with their forelegs with enough force to kill a person. Although there is a real danger in approaching any wild animal too closely, reports that White-tailed Deer are responsible for far more human fatalities annually than all North American bears misrepresent their demeanor. While true, these statistics include human fatalities resulting from vehicle

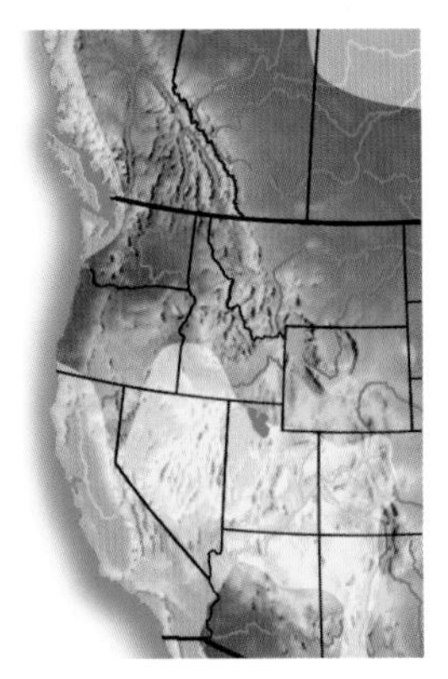

RANGE: From the southern third of Canada, the White-tailed Deer ranges south into the northern quarter of South America. It is largely absent from Nevada, Utah and California. It has been introduced to New Zealand, Finland, Prince Edward Island and Anticosti Island.

DID YOU KNOW?

The White-tailed Deer is named for the bright white undersides of its tail. A deer raises, or "flags," its tail when it is alarmed. The white flash of the tail communicates danger to nearby deer and provides a guiding signal for following individuals.

Total Length: 4½–7 ft (1.4–2.1 m)

Shoulder Height: 27–45 in (69–114 cm)

Tail Length: 8¼–14 in (21–36 cm)

Weight: 110–440 lb (50–200 kg)

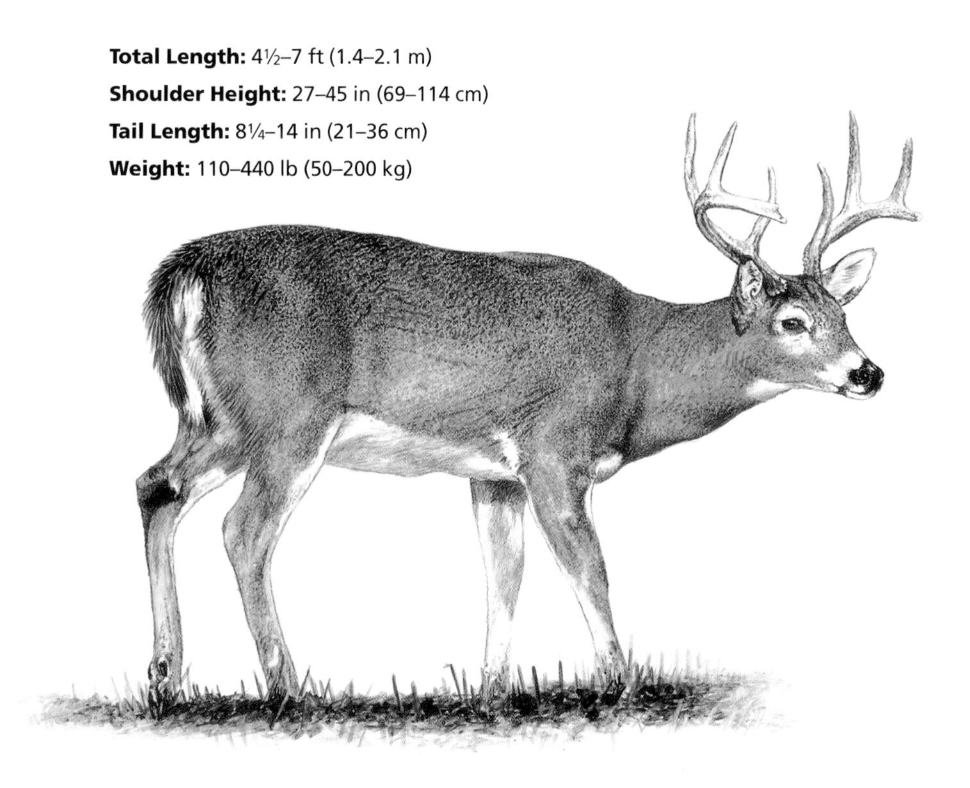

collisions with deer. Each year, several hundred thousand deer are involved in collisions on North American roads and highways.

ALSO CALLED: Flag-tailed Deer.

DESCRIPTION: The upperparts are generally reddish brown in summer and grayish brown in winter. The belly, throat, chin and underside of the tail are white. There is a narrow white ring around the eye and a band around the muzzle. A buck starts growing antlers in his second year. The antlers first appear as unbranched "spikehorns"; later, generally unbranched tines grow off the main beam. The main beams, when viewed from above, usually take the shape of a Valentine heart, with their origin a short distance above the notch of the heart and the terminal tines ending just before the apex. The metatarsal gland on the outside of the lower hindleg is about 1 in (2.5 cm) long.

HABITAT: The optimum habitat for a White-tailed Deer is rolling country with a mixture of open areas near cover. This deer frequents valleys and stream courses, woodlands, meadows and abandoned farmsteads with tangled shelterbelts. Areas cleared for roads, parking lots, summer homes, logging and mines support much of the vegetation on which the White-tailed Deer thrives.

FOOD: During winter, the leaves and twigs of evergreens, deciduous trees and brush make up most of the diet. In early spring and summer, the diet shifts to forbs, grasses and even mushrooms. On average, a White-tailed Deer eats 4½–11 lb (2–5 kg) of food a day.

DEN: A deer's bed is simply a shallow, oval, body-sized depression in leaves or snow. Favored bedding areas, which are often in secluded spots with good all-around visibility where deer can remain safe while they are inactive, will have an accumulation of new and old beds.

YOUNG: A doe gives birth to one to three fawns in late May or June, after a gestation of 6½ to 7 months. At birth, a fawn weighs about 6½ lb (2.9 kg), and its coat is tan with white spots. The fawn can stand and suckle shortly after birth, but it spends most of the first month lying quietly under the cover of vegetation. It is weaned when it is about four months old. A few well-nourished females may mate as fall fawns, but most wait until their second year.

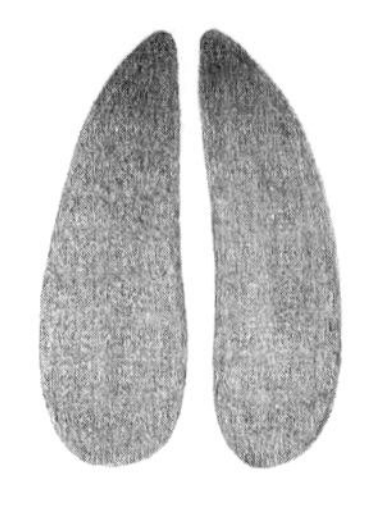

hoofprint (walking)

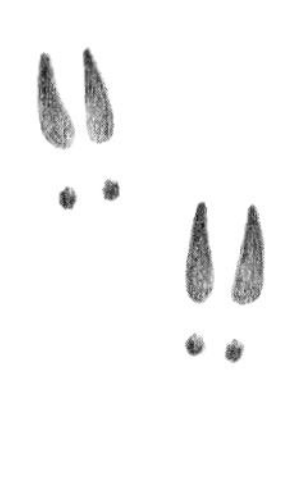

gallop group

SIMILAR SPECIES: The Mule Deer (p. 52) looks very similar, but it has a whitish rump patch and much longer metatarsal glands, and a buck's antlers usually have forked tines. The Pronghorn (p. 44) has black horns, a large white rump patch and white throat bands, and the lower half of its sides are white.

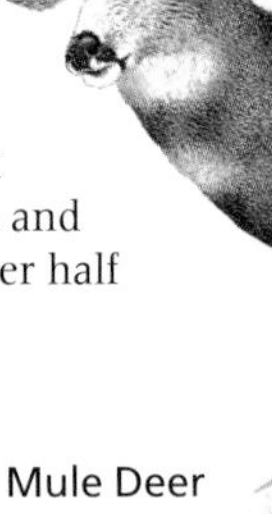

Mule Deer

Moose

Alces alces

The monarch of northern and mountain forests, the Moose is a handsome animal that provides a thrilling sight for tourists and wildlife enthusiasts. People who know it only from TV cartoon characterizations may not have such feelings for the Moose, but those who have followed its trails through waist-deep snow and mosquito-ridden bogs respect its abilities. The renowned Albertan mammologist, J. Dewey Soper, a man of the woods and admirer of Moose, wrote, "The peculiar, hoarse bellowing of the bull moose in the mating season is a memorable, far-reaching sound fraught with tingling qualities of the primordial. While deep-throated and raucous to the human ear, it doubtless broadcasts haunting and seductive overtones to the patiently waiting [cow moose] in the woods."

The Moose's long legs, short neck, humped shoulders and big, bulbous nose may lend it an awkward appearance, but they all serve it well in its environment. With its long legs, the Moose can easily step over downed logs and forest debris and cross streams. Deep snow, which seriously impedes the progress of wolves, is no obstacle for the Moose, which lifts its legs straight up and down to create very little snowdrag. The short neck holds the head, with its huge battery of upper and lower cheek teeth, in a perfect position for the Moose to nip off the twigs that make up most of its winter diet. The big bulbous nose and lips hold the twigs in place so the lower incisors can rip them off.

Winter ticks are often a problem for Moose. A single Moose can carry more than 200,000 ticks, and their irritation causes the moose to rub against trees for relief. With excessive rubbing, a Moose will lose much of its guard hair, resulting in the pale gray "ghost" moose that are sometimes seen in late winter. Winter Moose deaths are usually the result of blood loss to the ticks, rather than starvation—the twigs, buds and bark of deciduous trees and shrubs that form the bulk of its winter diet are rarely in short supply. The Moose's common name can also be traced to this feeding habit: the Algonquin called it *moz*, which means "twig eater." The summer diet of aquatic vegetation and other greenery seems quite palatable and varied, but even then, more than half the intake is wood.

DESCRIPTION: The Moose is the largest living deer in North America. The dark, rich brown to black upperparts fade to

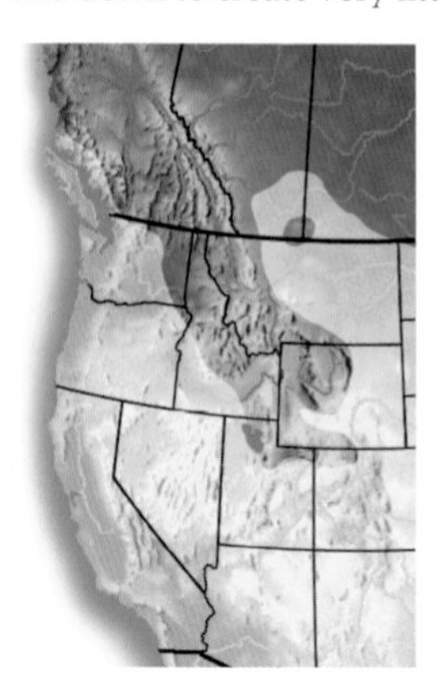

RANGE: In North America, this holarctic species ranges through most of Canada and Alaska. Its range has southward extensions through the Rocky Mountains, into the northern Midwest states and into New England and the northern Appalachians. The Moose is expanding into the farmlands of the northern Great Plains.

DID YOU KNOW?

The Moose is an impressive athlete: individuals have been known to run as fast as 35 mph (56 km/h), swim continuously for several hours, dive to depths of 20 ft (6.1 m) and remain submerged for up to a minute.

Total Length: 8–10 ft (2.4–3 m)
Shoulder Height: 5½–7 ft (1.7–2.1 m)
Tail Length: 3½–7½ in (8.9–19 cm)
Weight: 500–1180 lb (230–540 kg)

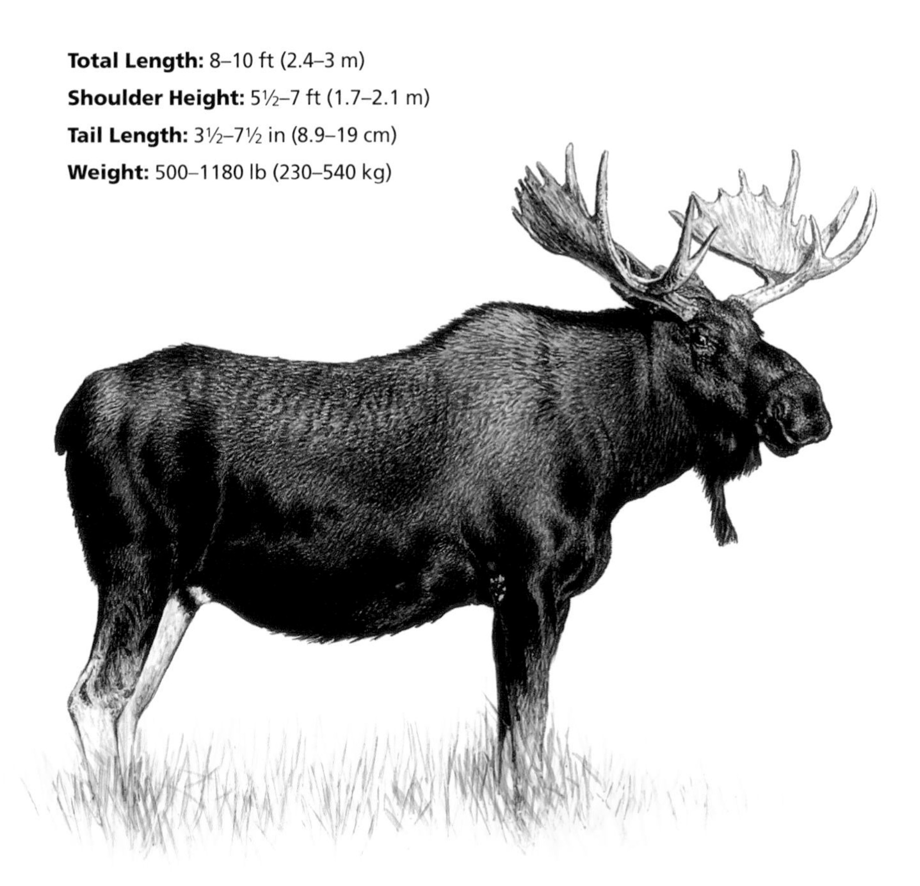

lighter, often grayish tones on the lower legs. The head is long and almost horse-like. It has a humped nose, and the upper lip markedly overhangs the lower lip. In winter, a mane of hair as long as 6 in (15 cm) develops along the spine over the humped shoulders and along the nape of the neck. In summer, the mane is much shorter. Both sexes usually have a large dewlap, or "bell," hanging from the throat. Only bull Moose have antlers. Unlike the antlers of other deer, the Moose's antlers emerge laterally, and many of the tines are usually merged throughout much of their length, giving the antler a shovel-like appearance. Elk-like antlers are common in young bulls (and they are the only type seen in Eurasian specimens today). A cow Moose has a distinct light patch around the vulva. A calf Moose is brownish to grayish red during its first summer.

HABITAT: Typically associated with the northern coniferous forest, the Moose is most numerous in the early successional stages of willows, balsam poplars and aspens. In less-forested foothills and lowlands, it frequents streamside or brushy areas with abundant deciduous woody plants. In summer, it may range well up into the subalpine or tundra areas of the mountains.

FOOD: About 80 percent of the Moose's diet is wood, mostly twigs and branches. In summer, it also feeds on submerged vegetation, sometimes sinking completely below the surface of a lake to acquire the succulent aquatics, but they never make up a large part of the diet. It prefers deciduous trees and shrubs over conifers.

DEN: The Moose makes its daytime bed in a sheltered area, much like other members of the deer family, and it leaves ovals of flattened grass from its weight. Other signs around the bed include tracks, droppings and browsed vegetation.

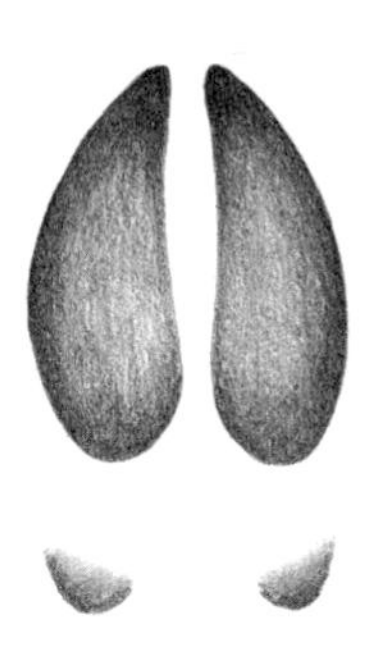

hoofprint

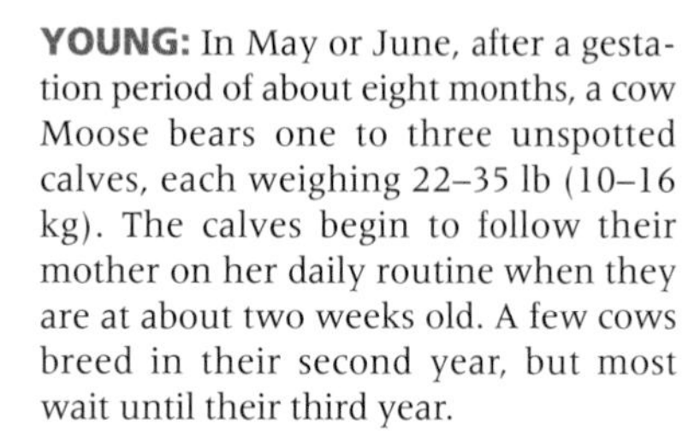

YOUNG: In May or June, after a gestation period of about eight months, a cow Moose bears one to three unspotted calves, each weighing 22–35 lb (10–16 kg). The calves begin to follow their mother on her daily routine when they are at about two weeks old. A few cows breed in their second year, but most wait until their third year.

trotting trail

SIMILAR SPECIES: With its large size and long head, the Moose resembles a bay or black Horse (p. 68) more than any native mammal. The Elk (p. 48) and the Caribou (p. 64) are both lighter in color, and the bulls do not have the lateral, palmate antlers of a bull Moose.

Horse

Caribou

Rangifer tarandus

The Caribou carves out a living in the deep snows and black fly fens where most other species of deer do not venture. It appears to do best in areas of expansive wilderness that allow it to undergo seasonal migrations between its summer and winter feeding grounds. This northern specialist is better adapted to cold climates than other deer—even the Caribou's nose is completely furred. The Caribou's winter coat has hollow guard hairs up to 4 in (10 cm) long that surmount a fine, fleecy, insulating undercoat. The hollow hairs provide excellent flotation (as well as insulation) when an animal is swimming across rivers and lakes during its lengthy migrations.

The Caribou's broad hooves are a great help in securing it a tasty meal, whether it has to walk high upon the snow to reach the old man's beard lichens hanging from the spruce trees or dig through snowpack to expose ground-dwelling cladonia lichens. The bristle-like hairs that cover a Caribou's feet in winter may help prevent the snow from abrading its skin when the Caribou digs feeding craters. This feeding strategy has been one of the Caribou's best-known characteristics for centuries—its name comes from eastern Canada, from the Micmac name *halibu*, which means "pawer" or "scratcher."

Unlike all other North American cervids, both sexes of the Caribou grow antlers. Not all Caribou shed their antlers at the same time: mature bulls shed their large sweeping racks in December; younger bulls retain theirs until February; cows keeps theirs until April (within a month they are growing a new set). After losing their antlers, the bulls become subordinate to the still-antlered cows, which are then better equipped to defend desirable feeding sites.

Antlers are but one of many characteristics that make this graceful wildland trotter distinctive and special. The fragmentation of the Rocky Mountain Caribou populations is of serious concern to resource managers, biologists and naturalists. There are few places where you can be assured of seeing this threatened animal, and seasonality greatly influences their whereabouts. In the Rocky Mountains, Caribou spend summer at high elevations to avoid the heat and the flies, and, in spring and fall, they migrate between the mountains and the foothill forests.

The seasonal migrations of the Rocky Mountain Caribou are more pronounced

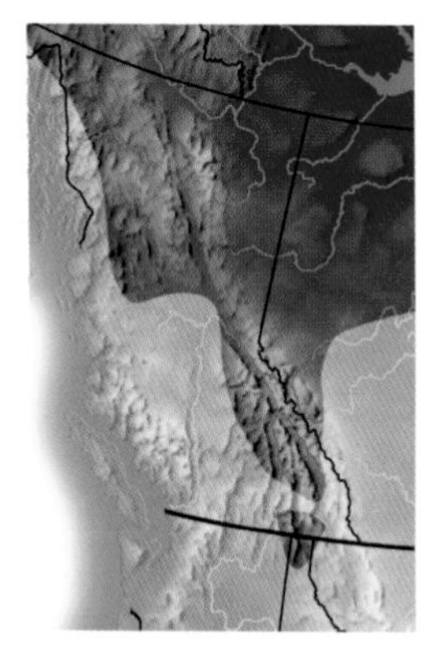

RANGE: The North American range of this holarctic animal covers most of Alaska and northern Canada, from the Arctic Islands south into the boreal forest. Its range has a southward extension through the Canadian Rockies and Columbia Mountains.

DID YOU KNOW?

Lichens, the Caribou's favorite winter food, grow very slowly and are often restricted to older spruce and fir forests, but a herd's erratic movements typically prevent it from overgrazing one particular area.

Total Length: 5½–8 ft (1.7–2.4 m)

Shoulder Height: 3–5½ ft (0.9–1.7 m)

Tail Length: 5–9 in (13–23 cm)

Weight: 200–240 lb (91–109 kg)

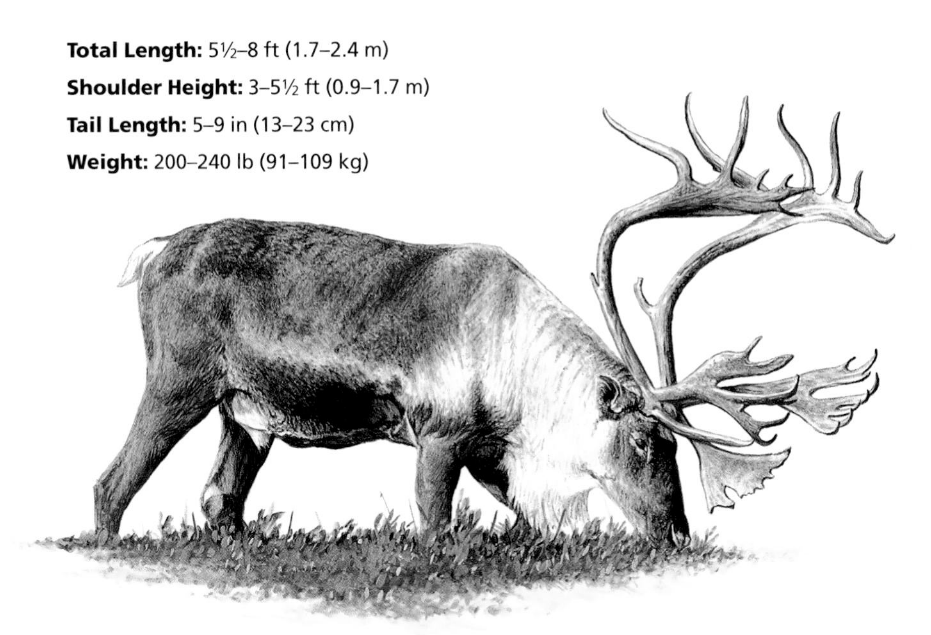

than the movements of the Caribou of the northern boreal forest, but they do not approach the incredible movements of Caribou in the Arctic. There was a time when the Rocky Mountain Caribou was considered one of four separate Caribou species in North America, but now all the North American Caribou and the Reindeer of Eurasia are grouped together as one species.

DESCRIPTION: In summer, the coat is brown or grayish brown above and lighter below, with white along the lower side of the tail and hoof edges. The winter coat is much lighter, with dark brown or grayish-brown areas on the upper part of the head, on the back and on the front of the limbs. Both sexes have antlers, but a bull's are much larger. Two tines come off the front of each main antler beam; one lower "brow" tine is palmate near the tip and is used to push snow to the side as the Caribou feeds. All the other distal tines come off the back of the main beam, which is a condition that is unique to the Caribou.

HABITAT: Most of the Caribou in the Rocky Mountains tend to remain in forests of spruce, fir, pine and aspen much of the year, but in summer they move into alpine meadows and the adjacent subalpine forest.

FOOD: Grasses, sedges, mosses, forbs, mushrooms and terrestrial and arboreal lichens make up the summer diet. In winter, the Caribou eats the buds, leaves and bark of both deciduous and evergreen shrubs, together with primarily arboreal lichens. This restless feeder takes only a few mouthfuls before walking ahead, pausing for a few more bites and then walking on again.

DEN: Like other cervids, the Caribou's bed is a simple, shallow, body-sized depression, often in a snowbank in summer. In winter, it usually lies with its body at right angles to the sun on exposed frozen lakes. Perhaps it absorbs more solar energy that way. Entire herds will sometimes lie in the same orientation.

YOUNG: Calving occurs in late May or June, after a gestation of about 7½ months. The unspotted calf, usually born singly (rarely as twins), weighs about 11 lb (5 kg) at birth. It often follows its mother within hours of birth, and it begins grazing when it is two weeks old. A calf may be weaned after a month, but some continue to nurse into winter. A cow usually first mates when she is 1½ years old; most males do not get a chance to mate until they are at least three to four years old.

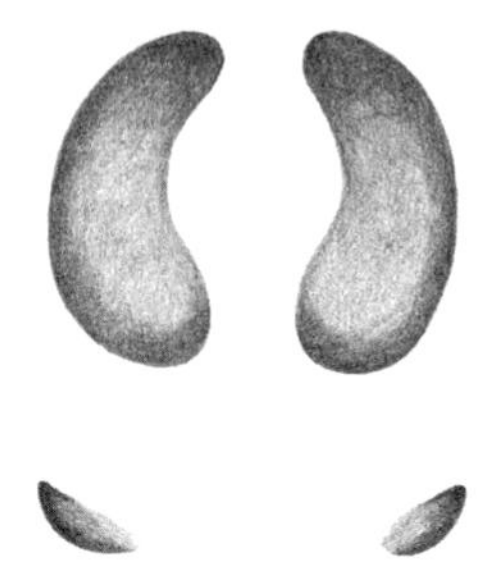

hoofprint

walking trail (in snow)

SIMILAR SPECIES: The heavy body and rectangular head of the Caribou distinguish it from the other members of the deer family, which have more triangular heads and less massive bodies. Both sexes bear antlers, and even calves may bear spikes, a feature that distinguishes them from other female or young deer. Also, a Caribou's feet produce clicking sounds when the animal is moving.

Elk

Horse

Equus caballus

Feral Horses in North America are descended from domesticated populations, and they have lived in the West for centuries. These wild Horses can usually be distinguished from their domestic kin by their much longer manes and tails and their patterns of behavior. Most of North America's wild Horses live in the Great Basin, but there are a few populations in the Rockies. The Pryor Mountain Wild Horse Range on the Montana-Wyoming border is one of the best places to see them.

Horse herds can have a different assortment of males and females, depending on the herd type. One type of herd is an accumulation of young bachelors. Males, usually over the age of two, that have left their parent herd may band together for a while, because no herd stallion will permit them near his mares. They stay together until either they find mares of their own or they are strong enough to steal mares from older stallions.

Another common type of herd is the mixed herd, in which a number of mares, a single adult stallion, a few young males and foals live and forage together. The mares in a herd like this are closely guarded by the stallion and are not free to come and go, but they are the ones that decide where the herd is going and what they are going to do each day.

In some harems there are two stallions, a number of subordinate mares and perhaps a few foals. In this grouping, the subdominant, usually younger, stallion exhibits "champing" behavior, in which he approaches the dominant stallion nose to nose with his ears forward in a gesture of friendly respect. The subdominant stallion is usually the offspring of one of the mares in the harem. As he matures and becomes the dominant stallion's equal, this champing behavior may become more threatening. Ultimately, a duel occurs. The two stallions face each other with their ears back, necks arched and tails high. They fight standing side-by-side, biting, kicking and pushing each other off balance. Eventually, one stallion is beaten and runs off, hopefully to find other mares to make a new harem. Mares of the harem stay together and accept the control of the victorious stallion. Only very rarely does a mare leave a harem to join a different herd.

Feral Horses use their teeth to groom the mane, neck and withers of another Horse, which helps develop and maintain the bonds between herd members.

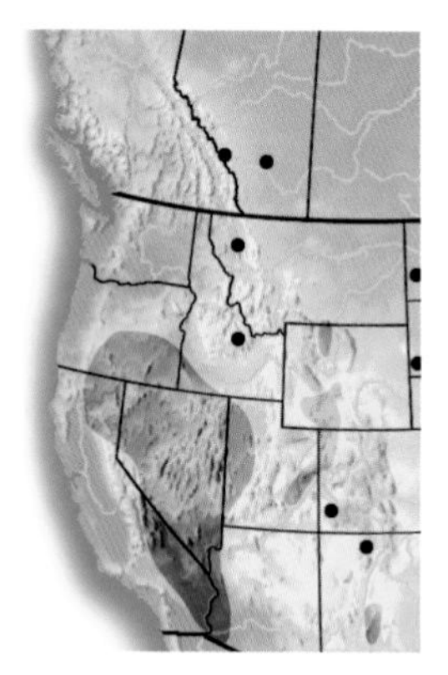

RANGE: Feral Horses occur in pockets along the Rocky Mountains from Alberta through Montana, Wyoming, Utah and Colorado. Much larger populations occur in the Great Basin, and there are other local herds in the southwest. Recently, efforts have been made to reduce their populations in many areas.

DID YOU KNOW?

Due to selective breeding, domestic Horse breeds vary greatly in size. The smallest are considerably less than 2 ft (61 cm) high at the shoulders, while the largest work horses are up to 6 ft (1.8 m) tall. Feral Horses are usually medium- to large-sized.

Total Length: up to 7 ft (2.1 m)

Shoulder Height: 3½–5½ ft (1.1–1.7 m)

Tail Length: up to 3 ft (91 cm)

Weight: 590–860 lb (270–390 kg)

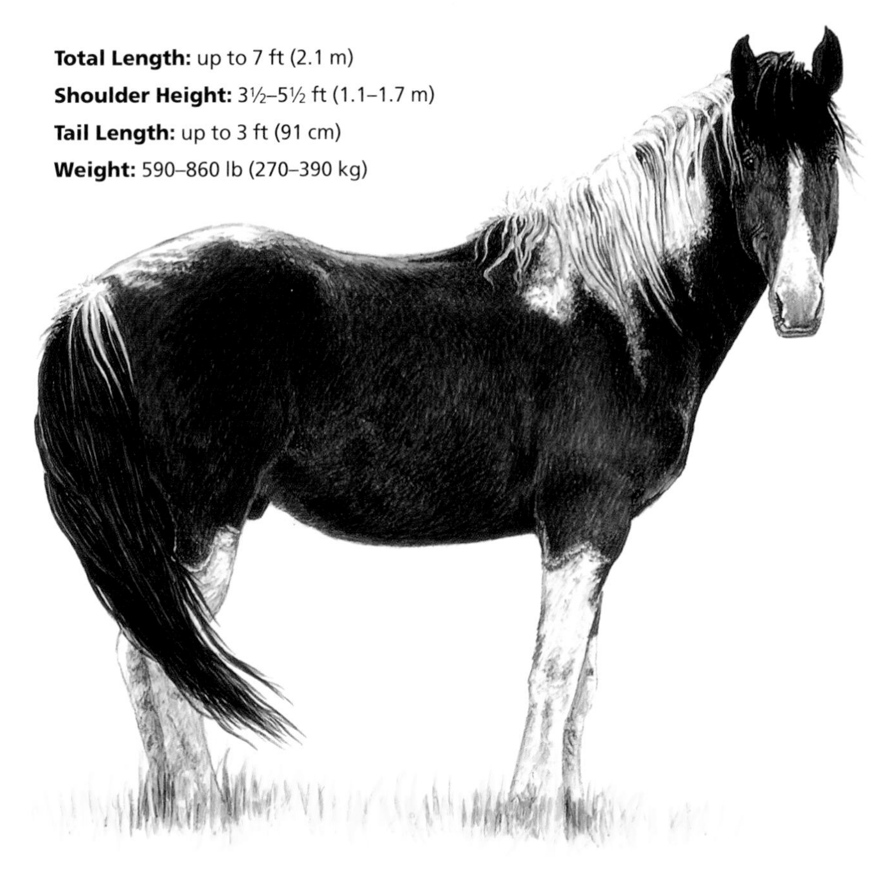

Biting flies seem to be a serious irritant to wild Horses, and after too many bites, a Horse may be in a state of extreme distress. To rid itself of the flies, the Horse may walk into thick foliage to scrape the flies off, roll mud to cover and soothe the skin or submerge itself in water.

ALSO CALLED: Mustang.

DESCRIPTION: Because of their domestic roots, wild Horses are extremely variable in size and color. They may be a solid color ranging from black to white or they may be spotted or bay or have various other color patterns. There are often white markings on the face, such as a star or blaze. They generally have a long mane and a long tail. Their hooves are semi-circular and uncloven.

HABITAT: Horses prefer areas of abundant vegetation beside watercourses, but they are so adaptable that they may be found from deserts to alpine tundra. In the Rocky Mountains, they are found in woodland areas, foothills, dry ridges, brushlands, and even marshy plateaus.

FOOD: As grazers, Horses spend as much as 80 percent of daytime hours grazing. Even at night they sleep only about 50 percent of the time—the rest of the night they are still grazing. Horses are herbivores, and they consume mainly grasses and forbs during the summer months. In winter, they eat woodier vegetation, such as the twigs or the bark of shrubs.

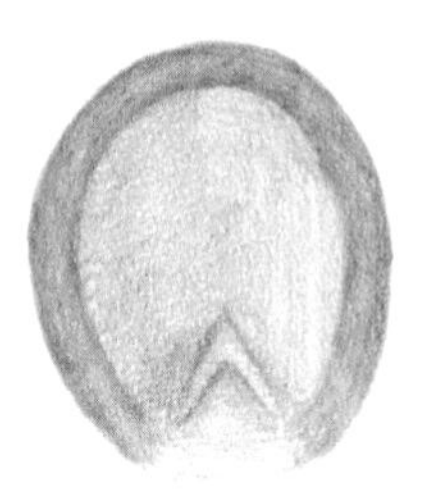

hoofprint

DEN: Feral Horses make no den for sleeping, but after lying down, a "bed" is visible where the grass was flattened. Although Horses are able to lie down, they often sleep standing up. While standing, the Horse closes its eyes and falls asleep. As soon as it slips into sleep, a highly specialized tendon in each leg locks the knee and prevents the leg from collapsing. As soon as the Horse wakes, the tendon is released and it can move. This mechanism evolved as a defensive strategy. As soon as the animal detects danger and is awakened, it is able to run, rather than having to take the time to rise from a prone position.

walking trail

YOUNG: A mare gives birth to one foal a year after a gestation of 11 months. Mating may occur during spring, summer or fall, often just a few days after a mare delivers her foal. The foal is precocious, and within a few hours of birth it is able to run with its mother and the rest of the herd. The first foal is weaned shortly before the next foal is born. If a mare is weak, or if food is scarce, she may have a foal only every second year.

SIMILAR SPECIES: Although large dark-colored Horses may be mistaken for Moose (p. 60) or Elk (p. 48) from a distance, the resemblance is restricted to the general size and color. No other animal has the same combination of a long-haired mane and tail, a single, uncloven hoof on each foot, and an elongated snout.

Moose

CARNIVORES

This group of mammals is aptly named, because, while some of its members are actually omnivorous (and eat a great deal of plant material), most of them will prey on other vertebrates. These "meat-eaters" vary greatly in size, and both the world's smallest member, the Least Weasel, and one of the largest, the Grizzly Bear, occur in the Rocky Mountains.

Cat Family (Felidae)

Canada Lynx

Excellent and usually solitary hunters, all our cats have long, curved, sharp, retractile claws. Like dogs, cats walk on their toes. They have five toes on each forefoot and four toes on each hindfoot, and their feet have naked pads and furry soles. As anyone who has a housecat knows, the top of a cat's tongue is rough with spiny, hard, backward-pointing papillae, which are useful to the cat for grooming its fur. Together with weasels, cats are among the most carnivorous of mammals.

Skunk Family (Mephitidae)

Biologists previously placed skunks in the weasel family, but recent investigations, including examinations of genetic sequences, have led taxonomists to group the North American skunks (together with the stink badgers of Asia) in a separate family. Unlike most weasels, skunks are usually boldly marked, and when threatened they can spray a foul-smelling musk from their anal glands.

Striped Skunk

Weasel Family (Mustelidae)

All weasels are lithe predators with short legs and elongated bodies. They have anal scent glands that produce an unpleasant-smelling musk, but, unlike skunks, they use it to mark territories rather than in defense. Most species are trapped for their valuable, long-wearing fur.

Wolverine

Raccoon Family (Procyonidae)

Common Raccoon

Raccoons and Ringtails are small to mid-sized carnivores that, like bears (and humans), walk on their heels. They are good climbers. Their teeth are adapted to eating vegetation as well as meat. They are best known for their long, banded, bushy tails. The Common Raccoon, of course, also has a distinctive, black facial mask.

Bear Family (Ursidae)

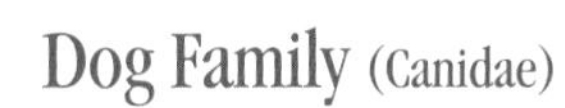
Black Bear

The three North American members of this family (two of which occur in the Rocky Mountains) are the world's largest terrestrial carnivores. All bears are plantigrade—they walk on their heels—and they have powerfully built forelegs and a short tail. Although most bears sleep through the harshest part of winter, they do not truly hibernate—their sleep is not deep and their temperature only drops a couple of degrees.

Dog Family (Canidae)

This family of dogs, wolves, coyotes and foxes is one of the most widespread terrestrial, non-flying mammalian families. The typically long snout houses a complex series of bones associated with the sense of smell, which plays a major role in finding prey and in intraspecific communications. Members of this family walk on their toes, and their claws are blunt and non-retractile.

Gray Wolf

Mountain Lion

Puma concolor

A pug-mark in the snow or a heavily clawed tree trunk are two powerful reminders that some places in the Rocky Mountains are still wild enough for the Mountain Lion. These large cats were once found throughout much of North America, but conflicts with settlers and their stock animals resulted in widespread removal of this great feline. Still, it is one of the most widespread, if not abundant, carnivores in both North and South America. Its alternate common names reflect this distribution: the name "puma" is derived from the name used for this animal by the Incas of Peru; "cougar" comes from Brazil.

The Mountain Lion is generally a solitary hunter, except when a mother is accompanied by her young. When the young are old enough, they follow their mother and sometimes even help her kill—a process that teaches the young how to hunt for themselves. Although Mountain Lions are capable of great bursts of speed and giant bounds, they often opt for a less energy-intensive hunting strategy. Silently and nearly motionless, a Mountain Lion waits in ambush in a tree or on a ledge until prey approaches. By leaping onto the shoulders of its prey and biting deep into the back of the neck while attempting to knock the prey off balance, the Mountain Lion can take down an animal as large as an adult Elk or a small Moose.

These big cats need the equivalent of about one deer a week to survive, and Mountain Lion densities in the Rocky Mountains tend to correlate with deer densities. Mountain Lions are adaptable creatures that may hunt by day or night. Hunting by day is quite common in the wilderness, but in areas close to human development they are active only at night.

The Mountain Lion, being one of the most charismatic animals of the Rocky Mountains, is a creature that every person hopes to see . . . from a safe distance. This elusive cat is a master of living in the shadows, but if you spend enough time hiking in the Rockies, you may one day see a streak of burnished brown flash through your peripheral vision. If it actually was a Mountain Lion, you can count yourself among the extremely lucky. Few people—even biologists— ever get more than a fleeting glimpse of these graceful felines. If you startle a Mountain Lion, which is quite improbable—it usually knows of your presence long before you know of its—it will quickly disappear from

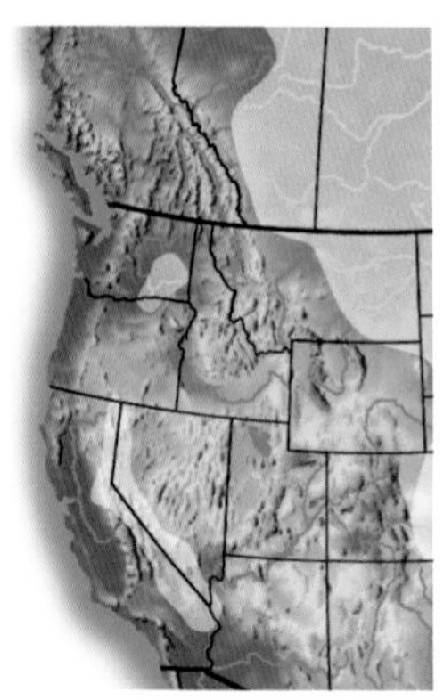

RANGE: The Mountain Lion formerly ranged from northern British Columbia east to the Atlantic and south to Patagonia. In North America, it has been extirpated from most areas except for the western mountains. A tiny population remains in the Everglades, and there are occasional reports from Maine and New Brunswick.

DID YOU KNOW?

During an extremely cold winter, a Mountain Lion can starve if the carcasses of its prey freeze solid before it can get more than one meal. This cat's jaws are designed for slicing, and it has trouble chewing frozen meat.

Total Length: 5–9 ft (1.5–2.7 m)

Shoulder Height: 26–32 in (66–81 cm)

Tail Length: 20–35 in (51–89 cm)

Weight: 70–190 lb (32–86 kg)

sight. Only young ones may come for a closer look at you. Young Mountain Lions, like most young carnivores, are extremely curious, and they don't yet realize that humans are best avoided.

ALSO CALLED: Cougar, Puma.

DESCRIPTION: This handsome feline is the only large, long-tailed native cat in the Rockies. It body is mainly buffy gray to tawny or cinnamon in color. Its undersides are pale buff or nearly white. It body is long and lithe, and its tail is more than half the length of the head and body. Its head, ears and muzzle are all rounded. The tip of the tail, the sides of the muzzle and the backs of the ears are black. Some individuals have prominent facial patterns of black, brown, cinnamon and white.

HABITAT: Mountain Lions are found most frequently in remote, wooded, rocky places, usually near an abundant supply of deer. In the Rocky Mountains, they inhabit mainly the montane regions, although sometimes they may venture into the subalpine, depending on food availability.

FOOD: In the Rockies, Mountain Lions rely mainly on deer. Other animals that they may also feed on include Bighorn Sheep, Mountain Goats, Elk, Moose, American Beavers and Common Porcupines. Even mice, rabbits, birds, domestic dogs and other cats may be consumed. In harsh winters, animals weakened by starvation may fall prey to Mountain Lions.

DEN: A cave or crevice between rocks usually serves as a den, but a Mountain Lion may also den under an overhanging bank, beneath the roots of a windthrown tree or even inside a hollow tree.

foreprint

YOUNG: A female Mountain Lion may give birth to a litter of one to six (usually two or three) kittens at any time of the year after a gestation of just over three months. The tan, black-spotted kittens are blind and helpless at birth, but their eyes open at two weeks. Their spots and mottled patterns help to camouflage them when their mother leaves to find food. As the kittens mature, they lose their spots and their rich blue eyes turn brown or hazel. They are weaned at about six weeks, by which time they weigh about 6½ lb (2.9 kg). Young Mountain Lions may stay with their mother for up to two years.

walking trail

SIMILAR SPECIES: The two other native cats in the Rocky Mountains, the Canada Lynx (p. 78) and the Bobcat (p. 82), are smaller (the Bobcat much so) and have mottled coats and bobbed tails.

Canada Lynx

Canada Lynx

Lynx canadensis

Meat is on the nightly menu for the Canada Lynx, and the meal of choice is the Snowshoe Hare. The classic predator-prey relationship of these two species is now well known to all students of zoology, but it took extensive field studies to determine how and why these two species interact to such a great extent. Periodic fluctuations in the numbers of Canada Lynx in local areas has been observed for decades: when hares are abundant, lynx kittens are more likely to survive and reproduce; when hares are scarce, many kittens starve and the lynx population declines, sometimes rapidly and usually one to two years after the decline in hares. Ernest Thompson Seton summed up the dependence of the Lynx on hares in his particular patter: "the Lynx lives on Rabbits, follows the Rabbits, thinks Rabbits, tastes . . . Rabbits, increases with them, and on their failure dies of starvation in the unrabbited woods."

This resolute carnivore copes well with the difficult conditions of its northern home. Its well-furred feet allow nearly silent movement and serve as snowshoes in deep winter snows. Like other cats, the Canada Lynx is not built for fast, long-distance running—it generally ambushes or silently stalks its prey. The ultimate capture of an animal relies on sheer surprise and a sudden overwhelming rush. With its long legs, it can travel rapidly while trailing evasive prey in the tight confines of a deep forest. It can also climb trees quickly to escape enemies or to find a suitable ambush site.

The reason why the Canada Lynx is so focused on the Snowshoe Hare as its primary prey may never be understood completely, but the forest community in which this cat lives certainly affects its lifestyle. Many other carnivores compete with the Canada Lynx for the same forest prey. Wolves, Coyotes, Red Foxes, Mountain Lions, Bobcats, Fishers, American Martens, Wolverines, American Minks, skunks, owls, eagles and hawks are all present in the same forests, and they all require animal prey for sustenance. Although these other predators may take a hare on occasion, none of them is as skilled at catching hares as the Canada Lynx.

The Canada Lynx is primarily a solitary hunter of remote forests. During peaks in its population, however, young cats may disperse into less hospitable environs. In recent memory, the Canada Lynx has been reported within the lim-

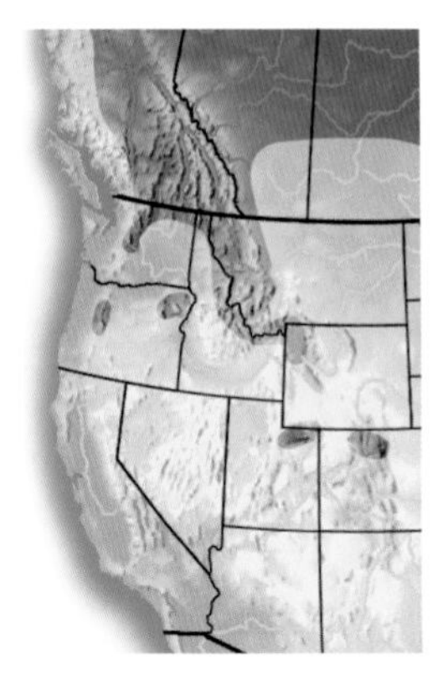

RANGE: Primarily an inhabitant of the boreal forest, the Canada Lynx occurs across much of Canada and Alaska. Its range extends south into the western U.S. mountains and into the northern parts of Wisconsin, Michigan, New York and New England.

DID YOU KNOW?

Some taxonomists consider the Canada Lynx to be the same species as the European Lynx (L. lynx), *which occupies the northern forests of Europe and Asia.*

Total Length: 31–40 in (79–102 cm)

Shoulder Height: 18–23 in (46–58 cm)

Tail Length: 3½–4¾ in (8.9–12 cm)

Weight: 15–40 lb (6.8–18 kg)

its of many major cities. These incidents are unusual, however, and the Canada Lynx typically avoids contact with humans. With each rare observation of a wild lynx, there undoubtedly comes a surprise to people who are well accustomed to the appearance of a house-cat—the stilt-legged lynx is more than twice the size and gangly in appearance.

DESCRIPTION: This medium-sized, short-tailed, long-legged cat has huge feet and protruding ears tipped with 2-in (5-cm) black hairs. The long, lax silvery-gray to buffy fur bears faint, darker stripes on the sides and chest and dark spots on the belly and insides of the forelegs. There are black stripes on the forehead and long facial ruff. The entire tip of the stubby tail is black. The long, buffy fur of the hindlegs makes it look like a lynx is wearing baggy trousers. Its large feet spread widely when it is walking, especially in deep snow. The footprint of this cat is wider than an adult human's hand.

HABITAT: The Canada Lynx is closely linked to northern coniferous forests. Numerous fallen trees and occasional dense thickets that serve as effective cover and ambush sites are desired habitat components. Lynx are dependent on their prey, and their prey are dependent on the twigs, grasses, leaves, bark and vegetation of the dense forest.

FOOD: The Snowshoe Hare typically makes up the bulk of the diet, but the Canada Lynx will sustain itself on squirrels, grouse, other rodents or even domestic animals. When lynx do not eat all of their kills, they cache the meat by covering it with snow or leafy debris.

foreprint

DEN: The den is typically an unimproved space beneath a fallen log, among rocks or even in a cave. Lynx do not share dens, and adult contact is restricted to mating. A mother lynx will share a den with her young until they are mature enough to leave.

YOUNG: Lynx breed in March or April, and the female gives birth to one to five (usually two or three) kittens in May or June. The kittens are generally gray, with indistinct longitudinal stripes and dark gray barring on the limbs. Their eyes open in about 12 days, and they are weaned at two months. They stay with their mother through the first winter and acquire their adult coats at 8 to 10 months. A female usually bears her first litter near her first birthday.

walking trail

SIMILAR SPECIES: The only other native cat that resembles the Canada Lynx is the smaller, shorter-legged Bobcat (p. 82). The Bobcat also has shorter ear tufts, and the tip of its tail is black above and white below.

Bobcat

Bobcat

Lynx rufus

For those of us who are naturalists as well as feline enthusiasts, our chances of seeing a Bobcat in the wild are much greater than seeing a Mountain Lion or Canada Lynx. Bobcats seem to be more tolerant of human presence, and they may even keep territories that border on developed land. Night drives through the southern Rockies may offer the best opportunities for seeing Bobcats, although, at best, the experience is a mere glimpse of this cat bobbing along in the headlights. Bobcats are also seen by many visitors to Yellowstone National Park.

The Bobcat looks like a large version of a housecat, but it has little of the housecat's domestication. A wildcat in every sense of the word, it impresses observers with its lightfootedness, agility and stealth, hopefully leaving the momentary experience forever retained within the viewer's mind.

Over the past two centuries, Bobcat populations have fluctuated greatly through their adaptability to human-wrought change and their vulnerability to our resentment. Less restricted in diet than the Canada Lynx, the Bobcat may vary its diet of hares with any number of small animals, including the occasional turkey or chicken . . . or two. Its farmyard raids did not go over well with the early settlers, and for more than 200 years the Bobcat was considered as vermin. Even today, this striking native feline remains on the "varmint" list in some states.

Despite its small size, the Bobcat is a ferocious and scrappy hunter that can take down animals much larger than itself. Reports from long ago that told of Bobcats killing deer were considered by the uninformed to be either tall tales or cases of mistaken identity. This remarkable feat, however, is indeed possible for a surreptitious Bobcat that waits motionless on a rock or ledge for a deer to approach. The Bobcat leaps onto the neck of the unsuspecting animal and then maneuvers to the lower side of the neck to deliver a suffocating bite to the deer's throat.

Bobcats may resort to such rough experiences in late winter when food is scarce, but they usually dine on simpler prey, such as rabbits, birds and rodents. Most of their prey, big or small, is caught at night in ambush. During the day, Bobcats remain immobile in any handy shelter.

Finding Bobcat tracks in soft ground may be the easiest way to determine the presence of this small cat in the moun-

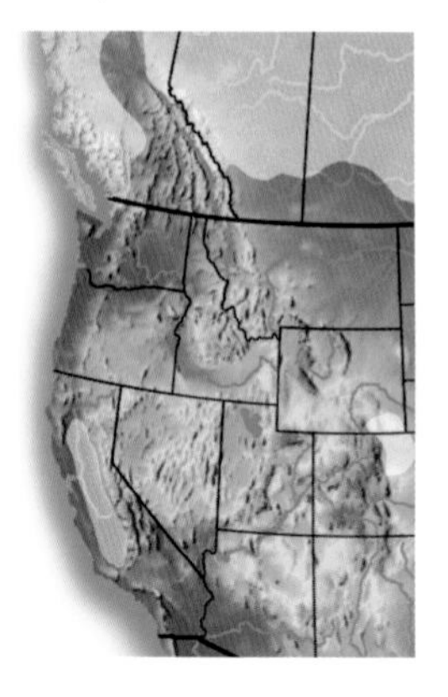

RANGE: The Bobcat has a spotty distribution across southern Canada and south to southern Mexico. It is scarce or absent in much of the Midwest.

DID YOU KNOW?

Most cats have long tails, which they lash out to the side to help them corner more rapidly in pursuit of prey. The Bobcat and the Canada Lynx, however, which typically hunt in brushy areas, have short, or "bobbed," tails that won't get caught in branches.

Total Length: 30–49 in (76–124 cm)

Shoulder Height: 17–21 in (43–53 cm)

Tail Length: 5–6¾ in (13–17 cm)

Weight: 9–40 lb (4.1–18 kg)

tains. Unlike Coyote or Red Fox prints, Bobcat prints rarely show any claw marks, and there is one cleft on the front part of the main foot pad and two on the rear. A Bobcat's print is quite like a large version of a housecat's, except that it tends to be found much farther from human structures. Like all cats, Bobcats bury their scat, and their scratches and scrapings can help confirm their presence.

DESCRIPTION: The coat is generally tawny or yellowish brown, although it varies with the seasons. The winter coat is usually dull gray, with faint patterns. In summer, the coat often has a reddish tinge to it (the source of the scientific name *rufus*). A Bobcat's sides are spotted with dark brown, and there are dark, horizontal stripes on the breast and the outsides of the legs. There are two black bars across each cheek and a brown forehead stripe. The ear tufts are less than 1 in (2.5 cm) long. The chin and throat are whitish, as is the underside of the bobbed tail. The upper surface of the tail is barred, and the tip of the tail is black above and light below.

HABITAT: The Bobcat occupies open coniferous and deciduous forests and brushy areas. It especially favors willow stands, which offer excellent cover for its clandestine hunting. Where the Canada Lynx is absent, it may range well up into mountain forests. It cannot tolerate too much snow, however, and is therefore deterred from expanding into more northerly regions.

FOOD: The preferred food seems to be rabbit, but the Bobcat will catch and eat squirrels, rats, mice, voles, beavers, skunks, wild turkeys and other ground-nesting birds. When necessary, it feeds on the kills of other animals, and it may even take down its own large prey, such as a deer or Pronghorn.

foreprint

DEN: Bobcats do not keep a permanent den. During the day, they use any available shelter. Female Bobcats prefer rocky crevices for the natal den, but they may also use hollow logs or the cavity under a fallen tree. The mothers do not provide a soft lining in the den for the kittens.

YOUNG: Bobcats typically breed in February or March, giving birth to one to seven (usually three) hairy, gray kittens in April or May, but they sometimes breed at other times of the year. The kittens' eyes open after nine days. They are weaned at two months, but they remain with the mother for three to five months. Female Bobcats become sexually mature at one year old; males at two.

walking trail

SIMILAR SPECIES: The Canada Lynx (p. 78) is the only other native, bob-tailed cat in North America. These two cats are nearly the same size, but the length of their hindlegs is very different, which makes the lynx appear taller. The lynx also has much longer ear tufts, and the tip of its tail is entirely black.

Canada Lynx

Western Spotted Skunk

Spilogale gracilis

To watch the antics of a Western Spotted Skunk as it prepares to spray is almost worth the putrid penalty. Almost. When agitated and in fear for its safety, the Western Spotted Skunk resorts to the practice that has made this family infamous. If foot stamping and tail raising does not convey sufficient warning, the next stage certainly will.

Unlike the more familiar Striped Skunk, which sprays in a U-position, with all its feet planted, the Spotted Skunk literally goes over the top in its spraying ways. Like a contortionist in a sideshow circus, this little skunk faces the threat and performs a handstand, letting its tail fall toward its head. The skunk can maintain this balancing act for more than five seconds, which is usually sufficient time to take aim and expel a well-placed stream of fetid scent into the face of the threat. Many animals may attempt to kill and eat this skunk before it sprays, but few are successful.

One of the first signs of spring in the wilds of the Rockies is the smell of skunk in the spring air. Motorists encountering this odor should not direct their disgust toward the Western Spotted Skunk, however, because it is an infrequent roadkill victim. Even though the Western Spotted Skunk is just about as numerous as the Striped Skunk in some parts of the Rockies, it is extremely agile and moves with the dexterity of its weasel relatives. With surprising ease, it can climb up to holes in hollow trees or to bird nests, where it finds food or shelter.

DESCRIPTION: This small skunk is mainly black with a white forehead spot and a series of four or more white stripes broken into dashes on the back. The pattern of white spots is different on each individual. The tail is covered with long, sparse hairs. The tip of the tail is white and the underside is black. The ears are small, rounded and black, and the face strongly resembles a weasel's. This skunk may walk, trot, gallop or make a series of weasel-like bounds. At night, its eyeshine is ale amber.

HABITAT: Western Spotted Skunks are found in woodlands, rocky areas, open prairies or scrublands. They do not occupy marshlands or wet areas, but farmlands make an excellent home. They are mainly nocturnal, and even in prime habitat they are seldom seen.

FOOD: This omnivorous mammal feeds on great numbers of insects, berries,

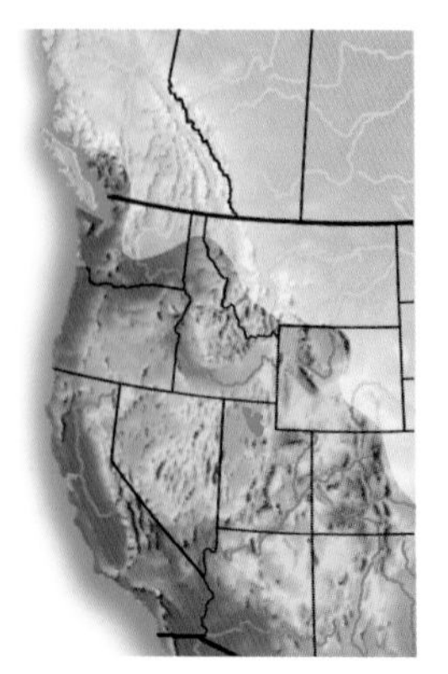

RANGE: The Western Spotted Skunk ranges from southwestern British Columbia to the southern tip of Texas and south through Mexico.

DID YOU KNOW?

Spotted skunks become sexually mature at a very young age. A male may be able to mate when he is just five months old, and a female usually mates in September or October of her first year.

Total Length: 9–18 in (23–46 cm)

Tail Length: 3⅜–7 in (8.6–18 cm)

Weight: 1½–2⅝ lb (0.7–1.2 kg)

eggs, nestling birds, small rodents, lizards and frogs. Animal matter usually accounts for the larger part of its diet. The Western Spotted Skunk is an opportunistic forager, and it will eat nearly anything that it finds or can catch. Insects, especially grasshoppers and crickets, are the most important food in summer, and small mammals are significant in fall and spring. It usually eats little or nothing in midwinter.

DEN: The Western Spotted Skunk is nomadic by comparison to the Striped Skunk. It rarely makes a permanent den, preferring to hole up temporarily in almost any safe spot: rock crevices, fallen logs, buildings, woodpiles, the abandoned burrows of other mammals and even tree cavities may all be inhabited by this skunk. The natal den is used for a longer period than other den spaces, and differs primarily by the grass and leaves with which the female lines the inside for comfort. In harsh winters, several skunks may den together to conserve energy and wait out inclement weather.

YOUNG: Two to six (usually four) young are born in May or June. The eyes and ears are closed at birth. The young skunks are covered with fine fur that betrays their future pattern. The eyes open after one month, and the young begin playing together at 36 days. By two months they are weaned. The family frequently stays together through fall, and it may overwinter in the same den, not dispersing until the following spring.

SIMILAR SPECIES: The Striped Skunk (p. 88) is the only other animal in the Rockies with a black-and-white coat. As its name suggests, however, its white markings are in broad stripes.

Striped Skunk

Mephitis mephitis

The famed warning colors of skunks are so effective that they communicate their message even to people who know little or nothing else about wildlife. This recognition is enhanced by the tendency of skunks to be involved in collisions on highways—slow-moving Striped Skunks, led by their noses, find foraging in roadside ditches dangerously tempting.

Butylmercaptan is the reason behind all the stink. Seven different sulfide-containing "active ingredients" have been identified in the musk, which not only smells, but also irritates. A distressed Striped Skunk will twist its body into a "U" prior to spraying, so that both its head and tail face the threat. If a skunk successfully targets the eyes, there is intense burning, copious tearing and sometimes a short period of blindness. The musk is also known to stimulate nausea in humans.

Despite all these good reasons to avoid close contact with the Striped Skunk, it is surprisingly tolerant of observation from a discreet distance, and watching a skunk can be very rewarding—its movements contrast with the hyperactive norm of its weasel cousins. The Striped Skunk's activity begins at sundown, when it emerges from its daytime hiding-place. It usually forages among shrubs, but it often enters open areas, where it can be seen with relative ease. The Striped Skunk is a clumsy and opportunistic predator that frequently digs shallow pits in search of meals. During Rocky Mountain winters, its activity is much reduced, and skunks spend the coldest periods in communal dens.

The only regular predator of the Striped Skunk is the Great Horned Owl. Lacking a highly developed sense of smell, this owl does not seem to mind the skunk's odor—nor do many other birds that commonly scavenge skunk roadkills.

DESCRIPTION: This cat-sized, black-and-white skunk is familiar to most people. Its basic color is glossy black. A narrow white stripe extends up the snout to above the eyes, and two white stripes begin at the nape of the neck, run back on either side of the midline and meet again at the base of the tail. The tail often has a continuation of the white bands ending in a white tip, but there is much variation in the amount and distribution of the white markings. The foreclaws are long and are used for digging. The hindclaws are short. A pair

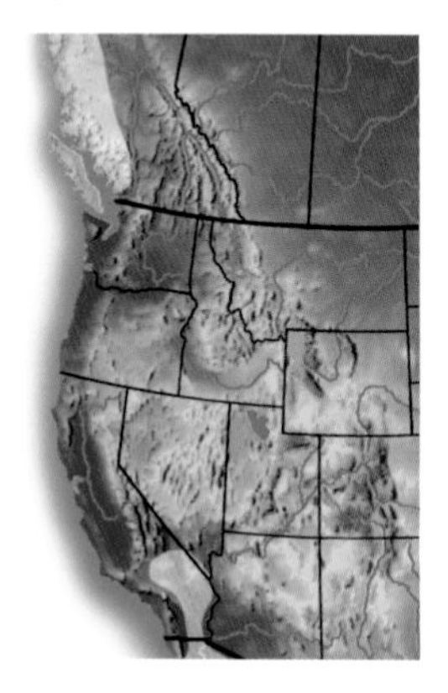

RANGE: The Striped Skunk is found across most of North America, from Nova Scotia to Florida in the east and from the southwestern Northwest Territories to northern Baja California in the west. It is absent from parts of the deserts of southern California.

DID YOU KNOW?

Fully armed, the Striped Skunk's scent glands have about 1 oz (30 ml) of noxious, smelly stink. The spray has a maximum range of almost 20 ft (6.1 m), and a skunk is accurate for half that distance.

Total Length: 21–31 in (53–79 cm)

Tail Length: 8⅝–12 in (22–30 cm)

Weight: 4¼–9¼ lb (1.9–4.2 kg)

of perineal musk glands on either side of the anus discharges the foul-smelling, yellowish liquid for which skunks are famous.

HABITAT: In the wild, the Striped Skunk seems to prefer streamside woodlands, groves of hardwood trees, semi-open areas, brushy grasslands and valleys. It also regularly occurs in cultivated areas, around farmsteads and even in the hearts of cities, where it is an urban nuisance that eats garbage and raids gardens.

FOOD: All skunks are omnivorous. Insects, including bees, grasshoppers, June bugs and various larvae, make up the largest portion, about 40 percent, of the spring and summer diet. To get at bees, skunks will scratch at a hive entrance until the bees emerge and then chew up and spit out great gobs of mashed bees, thus incurring the beekeeper's wrath. The rest of the diet is composed of fruits and berries, small mammals, bird eggs and nestlings, amphibians, reptiles, grains and green vegetation. Along roads, carrion is often an important component of a skunk's diet. By fall, small mammals, fruits and berries become more important in the diet.

DEN: In most instances, the Striped Skunk builds a bulky nest of dried leaves and grass in an underground burrow or beneath a building.

YOUNG: A female skunk gives birth to 2 to 10 (usually 5 or 6) blind, helpless young in April or May, after a gestation of 62 to 64 days. The typical black-and-white pattern of a skunk is present on the skin at birth. The eyes and ears open at three to four weeks. At five to six weeks, the musk glands are functional. Weaning follows at six to seven weeks. The mother and her young will forage together into the fall, and they often share a winter den.

SIMILAR SPECIES: Only the tiny Western Spotted Skunk (p. 86) also bears a black-and-white pattern, but its white areas are a series of spots or thin stripes, not the broad white stripes of the Striped Skunk. The American Badger (p. 108) has a white stripe running up its snout, but it is larger and squatter and has a grizzled, yellowish-gray, not black, body.

American Marten

Martes americana

Ferocity and playfulness are perfectly blended in the American Marten. This quick, active, agile weasel is equally at home on the forest floor or among branches and tree trunks. The fluidity of its motions and its attractive appearance juxtapose its carnivorous and often swift hunting tactics. A keen predator, the American Marten sniffs out voles, bird eggs and fledglings and acrobatically pursues Red Squirrels.

Unfortunately, this animal's playfulness, agility and insatiable curiosity are not easily observed, because it tends to inhabit wilder areas. The American Marten has been known to occupy human structures for short periods of time, should a food source be near, but more typical marten sightings are restricted to flashes across roadways or trails. The human pursuit of a marten rarely leads to a satisfying encounter—this weasel's mastery of the forest is ably demonstrated in its elusiveness.

The American Marten is a close relative of the Eurasian Sable (*M. zibellina*), and it is most widely known for its soft, lustrous fur. It is still targeted on traplines in remote wilderness areas.

As with so many species of forest mammals, marked fluctuations seem to occur every few years. Evidence from trappers and their records suggests that population levels of the American Marten are cyclical. Scientists tend to attribute these cycles to changes in prey abundance, whereas trappers often suggest that marten populations migrate from one area to another.

The American Marten is often used as an indicator of environmental conditions, because it is dependent upon food found in mature coniferous forests. Declining populations of this mustelid are indicators of a lack of food associated with the loss of mature forests. Hopefully, modern methods of logging and management will maintain American Marten populations and prevent the widespread declines that occurred previously.

ALSO CALLED: Pine Marten.

DESCRIPTION: This slender-bodied, fox-faced weasel has a beautiful, pale yellow to dark brown coat and a long, bushy tail. The feet are well furred and equipped with strong, non-retractile claws. The conspicuous ears are 1⅜–1¾ in (3.5–4.5 cm) long. The eyes are dark and almond shaped. The breast spot, when present, is usually orange but sometimes whitish or cream, and it varies in size from a small dot to a large

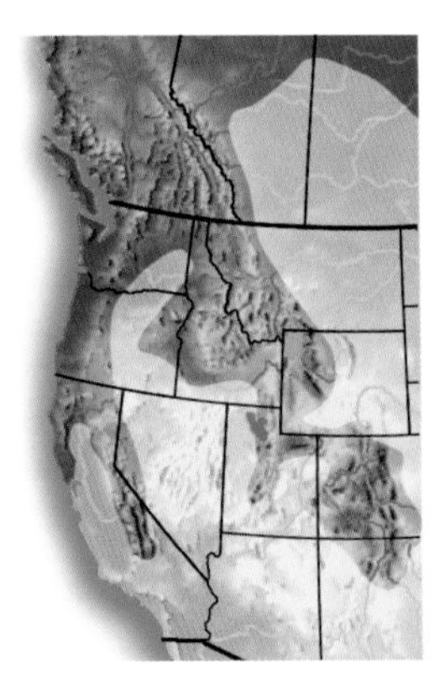

RANGE: The range of the American Marten coincides almost exactly with the distribution of boreal and montane coniferous forests across North America. It is re-establishing where mature forests have returned to areas that were formerly cut or burned.

DID YOU KNOW?

Although the American Marten, like most weasels, is keenly carnivorous, it has been known to consume an entire apple pie left cooling outside a window and to steal doughnuts off a picnic table.

Total Length: 19–26 in (48–66 cm)

Tail Length: 5½–7¾ in (14–20 cm)

Weight: 1¼–2¾ lb (0.6–1.2 kg)

patch that occupies the entire region from the chin to the belly. A male is about 15 percent larger than a female. There is a well-defined scent gland, about 3 in (7.6 cm) long and 1 in (2.5 cm) wide, on the center of the abdomen.

HABITAT: The American Marten prefers mature, particularly coniferous, forests that contain numerous dead trunks, branches and leaves to provide cover for its rodent prey. It does not occupy recently burned or cut-over areas.

FOOD: Although voles make up most of the diet, the American Marten is an opportunistic feeder that will eat squirrels, hares, bird eggs and chicks, insects, carrion and occasionally berries and other vegetation. In summer, it may enter the alpine tundra to hunt pikas and marmots. This active predator hunts both day and night.

DEN: The preferred den site is a hollow tree or log that the female lines with dry grass and leaves.

YOUNG: Breeding occurs in July or August, but with delayed implantation of the embryo, the litter of one to six (usually three or four) young isn't born until March or April. The young are blind and almost naked at birth and weigh just 1 oz (28 g). The eyes open at six to seven weeks, at which time the young are weaned from a diet of milk to one of mostly meat. The mother must quickly teach her young to hunt, because when they are only about three months old she will re-enter estrus, and, with mating activity, the family group disbands. Young female martens have their first litter at about the time of their second or third birthdays.

SIMILAR SPECIES: The Fisher (p. 92) is twice as long, seldom has an orange chest patch, has a long black tail and often has frosted or grizzled-gray to black fur. The American Mink (p. 102) has the white chin and irregular white spots on the chest, but it has shorter ears, shorter legs (it does not climb well) and a much less bushy, cylindrical tail.

Fisher

Martes pennanti

For the lucky naturalist, meeting a wild Fisher is a once-in-a-lifetime opportunity. For the rest of us, we must content ourselves with the knowledge that this reclusive animal remains a top predator in coniferous wildlands. Historically, the Fisher was more numerous, and it once ranged throughout the northern boreal forest, the northeastern hardwood forests and the forests of the Rocky Mountains and the Pacific ranges.

Fishers are animals of deep, untouched wilderness, and they often disappear shortly after development begins within their range. The clearing of forests, habitat destruction, fires and over-trapping resulted in its decline or extirpation over much of its range but there have been a few reintroductions and a gradual recovery in some areas over the past two decades.

The Fisher is one of the most formidable predator in the Rocky Mountains, and it could probably be considered the most athletic of this region's carnivores. Fishers are particularly nimble in trees, and the anatomy of their ankles allows the feet to rotate sufficiently that a Fisher can descend trees headfirst. Making full use of its athleticism during foraging, the Fisher incorporates any type of ecological community into its extensive home range, which can reach 75 mi (120 km) across. According to Ernest Thompson Seton, a well-known naturalist of the late 19th and early 20th centuries, as fast as a squirrel can run through the treetops, a marten can catch it, and as fast as the marten can run, a Fisher can catch and kill it.

The Fisher is a good swimmer, but, despite its name, it rarely eats fish. Perhaps this misnomer arose because of its confusion with the similar-looking American Mink, which does regularly feed on fish. These two weasels are most quickly distinguished by their preferred habitats: the American Mink inhabits riparian areas; Fishers prefer deep forests.

Few of the animals on which the Fisher preys can be considered easy picking; the most notable example is the Fisher's famed ability to hunt the Common Porcupine. What the porcupine lacks in mobility, it more than makes up for in defensive armor, and it requires all the Fisher's speed, strength and agility to yield a successful hunt. This hunting skill is far less common than wilderness tales suggest, however, and Fishers do not exclusively track porcupines; rather, they opportunisti-

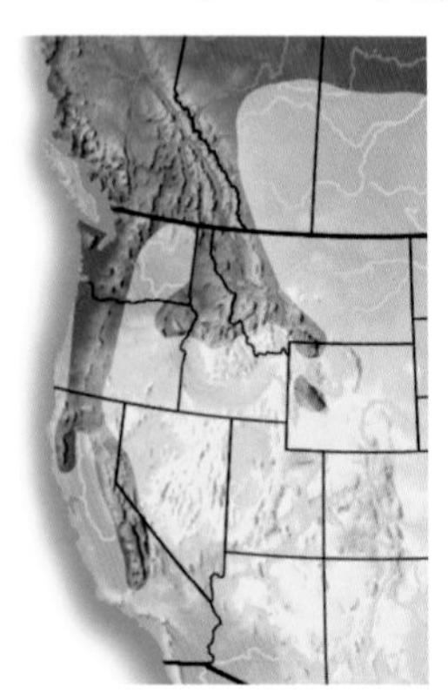

RANGE: Fishers occur across the southern half of Canada (except on the prairies) and range south through the Cascades-Sierra, through the Rockies to Wyoming and into New York and New England.

DID YOU KNOW?

The scientific name pennanti *honors Englishman Thomas Pennant. In the late 1700s, he predicted the decline of the American Bison and postulated that Native Americans entered North America via a Bering land bridge.*

Total Length: 31–40 in (79–102 cm)

Tail Length: 12–16 in (30–41 cm)

Weight: 3–13 lb (1.4–5.9 kg)

cally hunt whatever trail they cross. The majority of a Fisher's diet consists of rodents, rabbits, grouse and other small animals.

DESCRIPTION: The Fisher has a face that is fox-like, with rounded ears that are more noticeable than those on other large weasels. In profile, its snout appears distinctly pointed. The tail is dark and more than half as long as the body. The coloration over its back is variable, ranging from frosted gray to black. The undersides, tail and legs are dark brown. There may be a white chest spot. A male has a longer, coarser coat than a female, and he is typically 20 percent larger.

HABITAT: The preferred habitat is dense coniferous forests. It is not found in young forests or where logging or fire has thinned the trees. Fishers are most active at night and thus are seldom seen. They have extensive home ranges, and they may only visit a particular part of their range once every two to three weeks.

FOOD: Like other members of the weasel family, the Fisher is an opportunistic hunter, killing squirrels, hares, mice, muskrats, grouse and other birds. More than any other carnivore, however, the Fisher may hunt the Common Porcupine, which it kills by repeatedly attacking the head. It also eats berries and nuts, and carrion can be an important part of its diet.

DEN: Hollow trees and logs, brush piles, rock crevices, and cavities beneath boulders all serve as den sites. Most dens are only temporary lodging, because the Fisher is always on the move throughout its territory. The natal den is more permanent, and it is usually located in a safe place, such as a hollow tree. A Fisher may excavate its winter den in the snow.

YOUNG: A litter of one to four (usually two or three) young is born in March or April. The mother will breed again about a week after the litter is born, but implantation of the embryo is delayed until January of the following year. During mating, the male and female may remain coupled for up to four hours. The helpless young nurse and remain in the den for at least seven weeks, after which time their eyes open. When they are three months old, they begin to hunt with their mother, and by fall they are independent. The female is generally sexually mature when she is two years old.

left foreprint

walking trail

SIMILAR SPECIES: The American Marten (p. 90) is generally smaller and lighter in color, and it usually has a buff or orange chest spot. The American Mink (p. 102) is smaller and has shorter ears, shorter legs and a much less bushy, cylindrical tail. The Fisher typically has a more grizzled appearance than either the marten or mink.

American Mink

Short-tailed Weasel

Mustela erminea

Weasels have an image problem: they are often described as pointy-nosed villains and are frequently used to characterize dishonest cheats. These representations are unjust, however, because weasels are not manipulators, but rather earnest little predators with a flair for killing.

The Short-tailed Weasel is the Rocky Mountains' most common weasel, and it may be the most abundant land carnivore. Despite its abundance, the Short-tailed Weasel is not commonly seen, because, like all weasels, it tends to be most active at night and inhabits areas with heavy cover.

As Short-tailed Weasels roam about their ranges, they explore every hole, burrow, hollow log or brush pile for potential prey. In winter, they travel both above and below the snow in their search for prey. Once a likely meal is located, it is typically overwhelmed with a rush; then the weasel wraps its body around the animal and drives its needle-sharp canines into the back of the skull or neck. If the weasel catches an animal larger than itself, it seizes the prey by the neck and strangles it.

The Short-tailed Weasel's dramatic change between its winter and summer coats led Europeans to give it two different names: an animal wearing the dark summer coat is called a "stoat"; in the white winter pelage, it is known as an "ermine." In the Rocky Mountains, three weasel species alternate between white in winter and brown in summer, so the "stoat" and "ermine" labels are best avoided to prevent confusion.

DESCRIPTION: The short summer coat has brown upperparts and creamy white underparts, often suffused with lemon yellow. The last third of the tail is black. The short, oval ears extend noticeably above the elongated head. The eyes are black and beady. The long neck and narrow thorax make it appear as if the forelegs are positioned farther back than on most mammals and give the weasel a snake-like appearance. Starting in October and November, these animals become completely white, except for the black tail tip. The lower belly and inner hindlegs often retain the lemon yellow wash. In late March or April, the weasel molts back to its summer coat.

HABITAT: The Short-tailed Weasel is most abundant in coniferous or mixed forests and streamside woodlands. In summer, it may often be found in the

RANGE: In North America, this weasel occurs throughout most of Alaska and Canada and south to northern California and northern New Mexico in the West and northern Iowa and Pennsylvania in the East.

DID YOU KNOW?

These weasels typically mate in late summer, but after little more than a week the embryos stop developing. In early spring, up to eight months later, the embryos implant in the uterus and the young are born about one month later.

alpine tundra, where it hunts in rockslides and talus slopes.

FOOD: The diet appears to consist almost entirely of animals, including mice, voles, shrews, chipmunks, pocket gophers, pikas, rabbits, bird eggs and nestlings, insects and even amphibians. These weasels are quick, lithe and unrelenting in their pursuit of anything they can overpower. They often eat every part of a mouse except the filled stomach, which may be excised with surgical precision and left on a rock.

DEN: Short-tailed Weasels commonly take over the burrows and nests of mice, ground squirrels, chipmunks, pocket gophers or lemmings and modify them for weasel occupancy. They line the nest with dried grass, shredded leaves and the pelts and feathers of prey. Sometimes a weasel accumulates the pelts of so many animals that the nest grows to 8 in (20 cm) in diameter. Some nests are located in hollow logs, under buildings or in an abandoned cabin that once supported a sizable mouse population.

YOUNG: In April or May, the female gives birth to 4 to 12 (usually 6 to 9) blind, helpless young that weigh just 1⁄16 oz (1.8 g) each. Their eyes open at five weeks, and soon thereafter they accompany the adults on hunts. At about this time, a male has typically joined the family. In addition to training the young to hunt, he impregnates the mother and all her young females, which are sexually mature at two to three months. Young males do not mature until the next February or March, which is a reproductive strategy that reduces interbreeding between littermates.

SIMILAR SPECIES: The Least Weasel (p. 98) is generally smaller, and, although there may be a few black hairs at the end of the short tail, the entire tip is not black. The Long-tailed Weasel (p. 100) is generally larger and has orangish underparts, generally lighter upperparts and brown feet in summer.

summer colors

Total Length: 8¾–13 in (22–33 cm)

Tail Length: 1⅝–3½ in (4.1–8.9 cm)

Weight: 1⅝–3¾ oz (46–106 g)

Least Weasel

Mustela nivalis

If mice could talk, they would no doubt say that they live in constant fear of the Least Weasel. As it hunts, it enters and fully explores every hole it encounters. This pint-sized carnivore is small enough to squeeze into the burrows of mice and voles, and any small animal seems to warrant attack.

The Least Weasel is the smallest weasel (in fact, it is the smallest member of the carnivore order) in the world, but it has a monstrous appetite. On average, it consumes about 1/32 oz (1 g) of meat an hour, which means that it may eat almost its own weight in food in a day. If this weasel finds a group of mice or other small rodents, it is quick enough to kill them all within seconds. Prey that is not consumed immediately is stored to be eaten later.

Least Weasels can be active at any time of day, but they conduct most of their roaming at night. As a result, few people ever see these animals in action. Most human encounters with a Least Weasel result from lifting plywood, sheet metal or hay bales. These encounters are understandably brief—the weasel wastes little time in finding the nearest escape route, and any hole an inch across or greater is fair to enter, much to the dismay of its current resident.

Because the Least Weasel changes color with the seasons, a snowless fall or an early melt in spring can help make a weasel stand out against its environs. In spite of even this visual disadvantage, Least Weasels possess an uncanny ability to find shelter where there seems to be none.

DESCRIPTION: In summer, this small weasel is walnut brown above and white below. The short tail may have few black hairs at the end, but it never has an entirely black tip. The ears are short, scarcely extending above the fur. In winter, the entire coat is white, including the furred soles of the feet. Only a few black hairs may remain at the tip of the tail.

HABITAT: In the Rocky Mountains, the Least Weasel usually does not inhabit dense coniferous forests, preferring open grassy areas, forest edges or tundra instead. It sometimes occupies abandoned buildings and rock piles. Prey abundance seems to influence the distribution of Least Weasels more than habitat does.

FOOD: Voles, mice and insects are the usual prey, but amphibians, birds and eggs are taken when they are encountered.

RANGE: In North America, the Least Weasel's range extends from western Alaska through most of Canada and south to Nebraska and Tennessee. It is largely absent from southern Ontario, New York, New England and the Maritimes.

DID YOU KNOW?

During the fall molt, the white fur first appears on the animal's belly and spreads toward the back. The reverse occurs in spring: the brown coat begins to form on the weasel's back and moves toward its belly.

TOTAL LENGTH: 6–8¾ in (15–22 cm)

TAIL LENGTH: ⅞–1⅝ in (2.2–4.1 cm)

WEIGHT: ⅞–2⅝ oz (25–74 g)

summer colors

2x2 loping trail

DEN: The burrow and nest of a vole that fell prey to a Least Weasel makes a typical den site. The nest is usually lined with rodent fur and fine grass, which may become matted like felt and reach a thickness of 1 in (2.5 cm). In winter, frozen, stored mice may be dragged into the nest to thaw prior to consumption.

YOUNG: The Least Weasel does not exhibit delayed implantation of the embryos, and a female may give birth in any month of the year, after a gestation of 35 days. A litter contains 1 to 10 (usually 4 or 5) wrinkled, pink, hairless young. At three weeks they begin to eat meat. After their eyes open at 26 to 30 days, their mother begins to take them hunting. They disperse at about seven weeks old, living solitary existences except for brief mating encounters.

SIMILAR SPECIES: The Short-tailed Weasel (p. 96) is generally larger, has a longer tail with an entirely black tip and usually has a lemon yellow wash on the belly. The Long-tailed Weasel (p. 100) is larger, has a much longer, black-tipped tail and has orangish underparts, generally lighter upperparts and brown feet in summer.

left foreprint

Long-tailed Weasel

Mustela frenata

On a sunny winter day, there may be no better wildlife experience than to follow the tracks of a Long-tailed Weasel. This curious animal zig-zags as though it can never make up its mind which way to go, and every little thing it crosses seems to offer a momentary distraction. The Long-tailed Weasel seems continually excited, and this bountiful energy is easily read in its tracks as it leaps, bounds, walks and circles through its territory.

Long-tailed Weasels hunt wherever they can find prey: on and beneath the snow, along wetland edges, in burrows and even occasionally in trees. They can overpower smaller prey, such as mice, large insects and snakes, and kill them instantly. Larger prey species, up to the size of a rabbit, they grab by the throat and neck and wrestle to the ground. As the weasel wraps its snake-like body around its prey in an attempt to throw it off balance, it tries to kill the animal with bites to the back of the neck and head.

Unlike the Short-tailed Weasel and the Least Weasel, the Long-tailed Weasel only occurs in North America. With the conversion of native prairies to farmland, the Long-tailed Weasel has declined to the point where it is now regarded as a species of concern in much of its range. Still, in some native pastures that teem with ground squirrels, the Long-tailed Weasel can be found bounding about during the daytime, continually hunting throughout its waking hours.

DESCRIPTION: The summer coat is a rich cinnamon brown on the upperparts and usually orangish or buffy on the underparts. The feet are brown in summer. The tail is half as long as the body, and the terminal quarter is black. The winter coat is entirely white, except for the black tail tip and sometimes an orangish wash on the belly. As in all weasels, the body is long and slender—the forelegs appear to be positioned well back on the body—and the head is hardly wider than the neck.

HABITAT: The Long-tailed Weasel is an animal of open country. It may be found in agricultural areas, on grassy slopes and in the alpine tundra. Sometimes, in places where the Short-tailed Weasel is rare or absent, it forages in aspen parklands, intermontane valleys and open forests.

FOOD: Although the Long-tailed Weasel can successfully subdue larger prey than

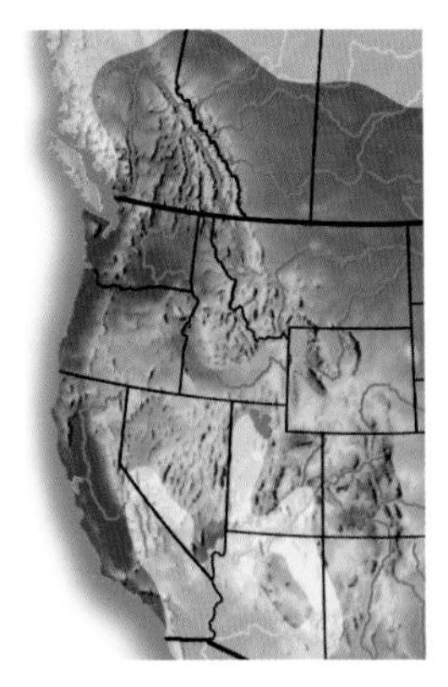

RANGE: From a northern limit in central British Columbia and Alberta, this weasel ranges south through most of the U.S. (except the southwestern deserts) and Mexico into northern South America.

DID YOU KNOW?

Weasel signs are not uncommon if you know what to look for: the tracks typically follow a paired pattern; the droppings are about the size of your pinkie finger, are twisted and full of hair, and are often left atop a rock pile.

Total Length: 13–19 in (33–48 cm)

Tail Length: 4¾–7½ in (12–19 cm)

Weight: 3–14 oz (85–400 g)

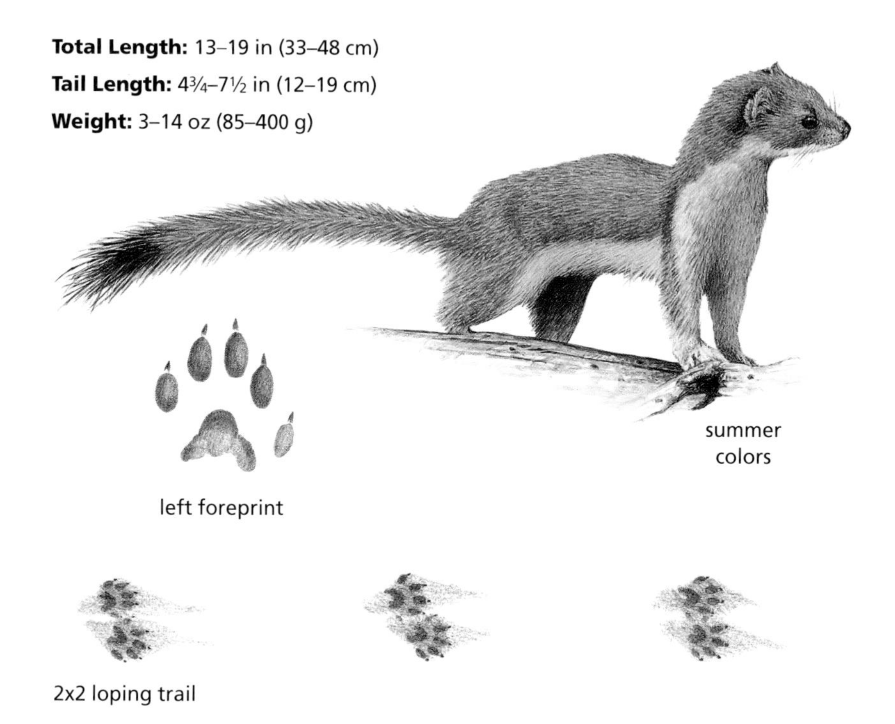

its smaller relatives, voles and mice still make up the majority of its diet. It also preys on ground squirrels, woodrats, Red Squirrels, rabbits and shrews, and it takes the eggs and young of ground-nesting birds when it encounters them.

DEN: The female usually makes her nest in the burrow of a ground squirrel or mouse that she has eaten. She often lines the nest with the fur or feathers of her prey.

YOUNG: Long-tailed Weasels typically mate in midsummer, but, through delayed implantation of the embryos, the young aren't born until April or May. The litter contains four to nine (usually six to eight) blind, helpless young. They are born with sparse white hair, which becomes a fuzzy coat by one week and a sleek coat in two weeks. At 3½ weeks the young begin to supplement their milk diet with meat; they are weaned when their eyes open, just after seven weeks. By six weeks, there is a pronounced difference in size, with young males weighing about 3½ oz (99 g) and females 2¾ oz (78 g). At about this time, a mature male weasel typically joins the group to breed with the mother and the young females as they become sexually mature. The group travels together and the male and female teach the young to hunt. The group disperses when the young are 2½ to 3 months old.

SIMILAR SPECIES: The Short-tailed Weasel (p. 96) is typically smaller, has a relatively shorter tail and has a lemon yellow (not orangish) belly and white feet in summer. The Least Weasel (p. 98) is much smaller and may have a few black hairs on the tip of its very short tail, but never an entirely black tip.

American Mink

Mustela vison

To many people, the liquid undulations of a bounding mink are more valuable than its much-prized fur. The American Mink is a smooth-traveling weasel that was described by naturalist Andy Russel as "moving ... like a brown silk ribbon." Indeed, like most weasels, the mink seems to move with the unpredictable flexibility of a toy slinky in a child's hands.

Minks are tenacious hunters, following scent trails left by potential prey over all kinds of obstacles and terrain. Almost as aquatic as otters, these opportunistic feeders routinely dive to depths of several yards in pursuit of fish. Their fishing activity tends to coincide with breeding aggregations of fish in spring and fall, or during winter, when low oxygen levels force fish to congregate in oxygenated areas. It is along watercourses, therefore, that minks are most frequently observed, and their home ranges often stretch out in linear fashion, following rivers for up to 3 mi (4.8 km).

The American Mink is active throughout the year, and it is often easiest to follow by trailing its winter tracks in snow. The paired prints left by its loping gait traces the inquisitive animal's adventures as it comes within sniffing distance of every burrow, hollow log and bush pile. This active forager always seems to be on the hunt; scarcely any feeding opportunities are passed up. Minks may kill more than they can eat, and surplus kills are stored for later use. A mink's food caches are often tucked away in its overnight dens, which are typically dug into riverbanks, beneath rock piles or in the mound of a permanently evicted muskrat.

DESCRIPTION: The sleek coat is generally dark brown to black, usually with white spots on the chin, chest and sometimes the belly. The legs are short. The tail is cylindrical and only somewhat bushy. A male is nearly twice as large as a female. The anal scent glands produce a rank, skunk-like odor.

HABITAT: The American Mink is almost never found far from water. It frequents wet zones in coniferous or hardwood forest, bushlands and streamside vegetation in the foothills.

FOOD: Minks are fierce predators of muskrats, but in their desire for nearly any meat they also take frogs, fish, waterfowl and their eggs, mice, voles, rabbits, snakes and even crayfish and other aquatic invertebrates.

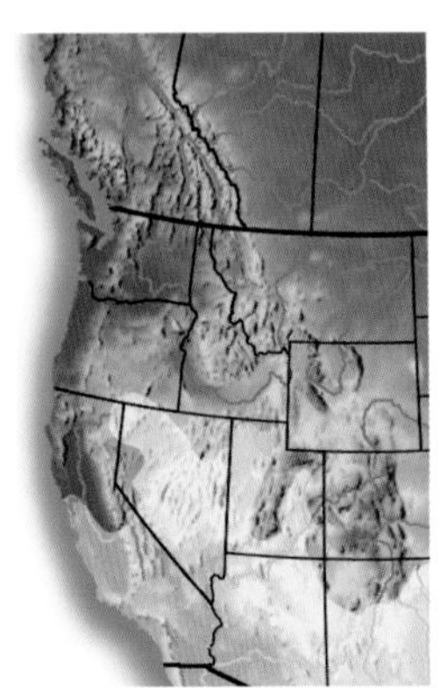

RANGE: This wide-ranging weasel occurs across most of Canada and the U.S., except for the high Arctic tundra and the dry Southwestern regions.

DID YOU KNOW?

"Mink" is from a Swedish word that means "stinky animal." Although not as aromatic as skunks, minks are the smelliest of the weasels. The anal musk glands can release the stinky liquid—but not aim the spray—when a mink is threatened.

Total Length: 17–24 in (43–61 cm)

Tail Length: 5–8¼ in (13–21 cm)

Weight: 1¾–5 lb (0.8–2.3 kg)

DEN: The den is usually in a burrow close to water. A mink may dig its own burrow, but more frequently it takes over a muskrat or beaver burrow and lines the nest with grass, feathers and other soft materials.

YOUNG: Minks breed any time between late January and early April, but because the period of delayed implantation is variable in length (from one week to 1½ months), the female almost always gives birth in late April or early May. The actual gestation period is about one month. There are 2 to 10 (usually 4 or 5) helpless, blind, pink, wrinkled young in a litter. The eyes open at 24 to 31 days. Weaning begins at five weeks. The mother teaches the young to hunt for two to three weeks, after which they fend for themselves.

SIMILAR SPECIES: The American Marten (p. 90) has a bushier tail, longer legs and an orange or buff throat patch, and it is not as sleek looking. The Northern River Otter (p. 112) is much larger and has a tapered tail and webbed feet.

Wolverine

Gulo gulo

The Wolverine is one of the most poorly understood mammals in North America. It is an elusive animal of deep wilderness, as well as a creature of many myths and tall tales. More recently, the Wolverine has become a symbol of deep, pristine wilderness. Although most of us will never see a Wolverine, the knowledge that it still maintains a hold in the northern forests may reassure us that expanses of wilderness still exist.

Tales of the Wolverine's gluttony—its reputation rivals that of hyenas in Africa—have lingered in forest lore for centuries. Pioneers warned their children against the dangers of the forests, and often they meant Wolverines. Why? Because the Wolverine is an efficient and agile predator: it can crush through bone in a single bite; it has long, semiretractile foreclaws that allow it to climb trees; and it is ferocious enough to challenge a lone bear or wolf. What we rarely hear about is this animal's intelligence, its uniqueness among its weasel relatives and its sheer vigor and beauty.

From the few behavioral studies of Wolverines, their character is perceived as less vicious and certainly more clever. Even simple observations of a Wolverine standing on its hindlegs and scanning the surroundings with a paw at its forehead to shield its eyes from the sun are indicative of behavior that we are only now starting to understand. Nevertheless, some of the Wolverine's reputation is well deserved. True to its nickname "skunk bear," the Wolverine produces a stink that rivals skunks in foulness. The abundant, stinky scent is produced in anal glands and is primarily used to mark territory.

The Wolverine's habitat preferences seem to vary as its diet shifts with seasons. In summer, it eats mostly ground squirrels, other small mammals, birds and berries; in winter it lives on carrion, mainly hoofed mammals, the majority of which it scavenges from wolves or roadkills. Like a vulture, the Wolverine has the ability to detect carcasses from far away. The largest weasel of all, the Wolverine has one of the mammal world's most powerful jaws, which it uses to tear meat off of frozen carcasses or to crunch through bone to get at the rich, nourishing marrow inside. Few other large animals are able to extract as much nourishment from a single carcass.

DESCRIPTION: Although the head is small and weasel-like, the long legs and

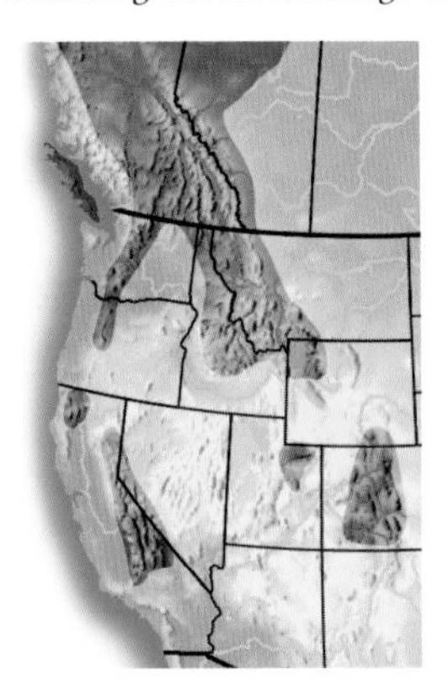

RANGE: In North America, the Wolverine is a species of the coniferous forests and tundra of Alaska and northern Canada. It follows the montane coniferous forests as far south as California and Colorado.

DID YOU KNOW?

The Wolverine's lower jaw is more tightly bound to its skull than most other mammals' jaws. The articulating hinge that connects the upper and lower jaws is wrapped by bone in adult Wolverines, and in order for the jaws to dislocate, this bone would have to break.

Total Length: 34–42 in (86–107 cm)

Tail Length: 7½–10 in (19–25 cm)

Weight: 15–35 lb (6.8–16 kg)

long fur look like they belong on a small bear. Unlike a bear, however, the Wolverine has an arched back and a long, bushy tail. The coat is mostly shiny, dark cinnamon brown to nearly black in color. There may be yellowish-white spots on the throat and chest. A buffy or pale brownish stripe runs down each side from the shoulder to the flank, where it becomes wider. These stripes meet just before the base of the tail, leaving a dark "saddle" on the back.

HABITAT: The Wolverine prefers large areas of remote wilderness, where it frequently occupies wooded foothills and mountains. In summer, it forages into the alpine tundra and hunts along slopes. In winter, it drops to lower elevations and may move far away from the mountains. The Wolverine's enormous territory encompasses a great variety of habitats. Wolverines are so agile and determined that likely no terrain is unconquerable to them.

FOOD: Wolverines prey on mice, ground squirrels, birds, beavers and fish. Deer, Caribou, Mountain Goats and even Moose have been attacked, often successfully. In winter, Wolverines often scavenge malnourished animals or the remains left by other predators. To a limited extent, they eat berries, fungi and nuts. Although Wolverines are generally thought to avoid human habitations, they are known to break into wilderness cabins and meat caches to eat or destroy everything within.

DEN: The den may be among the roots of a fallen tree, in a hollow tree butt, in a rocky crevice or even in a semi-permanent snowbank. The natal den is often underground, and it is lined with leaves by the mother. A Wolverine may maintain several dens throughout its territory, ranging in quality from makeshift cover under tree branches to a permanent underground dugout.

YOUNG: Wolverines breed between late April and early September, but the embryos do not implant in the uterine wall until January. Between late February and mid-April the female gives birth to a litter of one to five (generally two or three) cubs. The stubby-tailed cubs are born with their eyes and ears closed and with a fuzzy white coat that sets off the darker paws and face mask. They nurse for eight to nine weeks; then they leave the den and their mother teaches them to hunt. The mother and her young typically stay together through the first winter. The young disperse when they become sexually mature in spring.

loping trail

SIMILAR SPECIES: The American Badger (p. 108) is the only species you might mistake for a Wolverine, but a badger is squatter, has light patches on its head and does not have the lighter side stripes of a Wolverine. Older American Badgers often have the "wrap-around" jaw articulation seen in older Wolverines.

American Badger

American Badger

Taxidea taxus

Where grasslands rise up the Rocky Mountain slopes, American Badgers can be found. Badgers, with their flair for remodeling, are nature's rototillers and backhoes. They create the basis for many underground dens, shelters and hibernacula for a multitude of animals, both vertebrate and invertebrate. The large holes left by badgers are of critical importance as den sites for dozens of species, from Coyotes to Black-widow Spiders. When badgers are eliminated from an area, the populations of many of these burrow-dependent animals eventually decline.

The American Badger enjoys a reputation for fierceness and boldness that was acquired in part from a not-very-closely-related mammal bearing the same name in Europe. While it is true that a cornered badger will put up an impressive show of attitude—when it is severely threatened or in competition, the badger's claws, strong limbs and powerful jaws make this animal a dangerous opponent—like most animals it prefers to avoid a fight. In spite of its impressive arsenal, it is only ground squirrels and other small rodents that the badger routinely kills. However, rare occasions are known where badgers took Coyote pups. Conversely, a group of Coyotes may defeat a badger.

Pigeon-toed and short-legged, the American Badger is not much of a sprinter, but its heavy front claws enable it to move large quantities of earth in short order. Although a badger's predatory nature is of benefit to landowners, its natural digging skills have led many badgers to be killed—cattle and horses have been known (rarely) to break their legs when stepping carelessly into badger workings. Interestingly, this crippling misfortune never seems to be in evidence among native hoofed mammals.

Badgers tend to spend a great part of winter sleeping in their burrows, but they do not enter a full state of hibernation like their European relatives, or like the ground squirrels upon which they feed. Instead, badgers emerge from their slumber to hunt whenever winter temperatures are more moderate.

In spite of low population densities, almost all sexually mature female badgers are impregnated during the nearly three months that they are sexually receptive. As with most members of the weasel family, once the egg is fertilized further embryonic development is put off until the embryos implant, usually in January, which will result in a spring birth.

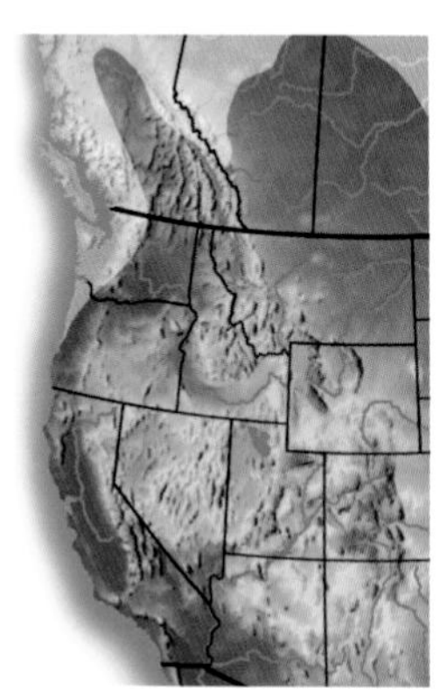

RANGE: From north-central Alberta and Saskatchewan, the American Badger ranges southeast throughout the Great Plains and prairies and southwest to Baja California and the central Mexican highlands.

DID YOU KNOW?

Badgers make an incredible variety of sounds: adults hiss, bark, scream and snarl; in play, young badgers grunt, squeal, bark, meow, chirr and snuffle; and the front claws clatter when a badger runs on a hard surface.

Total Length: 31–33 in (79–84 cm)

Tail Length: 5–6¼ in (13–16 cm)

Weight: 11–24 lb (5–11 kg)

DESCRIPTION: Long, grizzled, yellowish-gray hair covers these short-legged, muscular members of the weasel family. The hair is longer on the sides than on the back or belly, which adds to the flattened appearance of the body. A white stripe originates on the nose and runs back onto the shoulders or sometimes slightly beyond. The top of the head is otherwise dark. A dark, vertical crescent, like a badge, runs between the short, rounded, furred ears and the eyes. The sides of the face are whitish or very pale buff. The short, bottlebrush tail is more yellowish than the body. The lower legs and feet are very dark brown, becoming blackish at the extremities. The three central claws on each forefoot are greatly elongated for digging. The largest subspecies of badgers occur on the east slopes of the Rocky Mountains; those on the western side are much smaller.

HABITAT: Essentially an animal of open places, the badger shuns forests. It is usually found in association with ground squirrels, typically in the open grasslands of the parkland and prairies. In the mountains, it forages on treeless alpine slopes or in riparian meadows. It visits the alpine tundra in summer in search of marmots, pocket gophers and other burrowing prey.

FOOD: Burrowing mammals fulfill most of the badger's dietary needs, but it also eats eggs, young ground-nesting birds, mice and sometimes carrion, insects and snails.

DEN: A badger may dig its own den or take over a ground squirrel's burrow. The den may approach 30 ft (9.1 m) in length and have a diameter of about 1 ft (30 cm). It builds a bulky grass nest in an expanded chamber near or at the end of the burrow. A large pile of excavated earth is generally found to one side of the burrow entrance.

YOUNG: One to five (usually four) naked, helpless young are born between late April and mid-June. Their eyes open after a month, and at two months their mother teaches them to hunt. In early evening they leave the burrow, trailing their mother. The babies investigate every grasshopper or beetle they encounter, but the mother directs the expedition to ground squirrel burrows. She often cripples a ground squirrel and then leaves it for her young to kill. The young disperse in fall, when they are three-quarters grown. Some of the young females may mate in their first summer, but most badgers are not sexually mature until they are a year old. Delayed implantation of the embryo is characteristic.

left foreprint

walking trail

SIMILAR SPECIES:
The Wolverine (p. 104) is the only species that you might confuse with a badger, but the badger's body is much more flattened, and the white stripe on its nose is unique. Also, Wolverines lope, whereas badgers trot.

Wolverine

Northern River Otter

Lontra canadensis

It may seem to be too good to be true, but all those playful characterizations of the Northern River Otter are founded on truth. Otters often amuse themselves by rolling about, sliding, diving or "body surfing," and they may also push and balance floating sticks with their noses or drop and retrieve pebbles for minutes at a time. They seem particularly interested in playing on slippery surfaces—they leap onto the snow or mud with their forelegs folded close to their bodies for a streamlined toboggan ride. Unlike most members of the weasel family, river otters are social animals, and they will frolic together in the water and take turns sliding down banks.

With their streamlined bodies, rudder-like tails, webbed toes and valved ears and nostrils, river otters are well adapted for aquatic habitats. Even when they emerge from the water to clamber over rocks, there is a serpentine appearance to their progression. The large amounts of playtime they seem to have results from their efficiency at catching prey when it is plentiful. Although otters generally cruise along slowly in the water by paddling with all four feet, they can sprint after prey with the ease of a seal whenever hunger strikes. When an otter swims quickly, it chiefly propels itself with vertical undulations of its body, hindlegs and tail. Otters can hold their breath for as long as five minutes, and, if so inclined, they could swim the breadth of a small lake without surfacing.

Because of all their activity, Northern River Otters leave many signs of their presence when they occupy an area. Their slides are the most obvious and best-known evidence—but be careful not to mistake the slippery beaver trails that are common around beaver ponds for otter slides. Despite their other aquatic tendencies, otters always defecate on land. Their scat is simple to identify—it is almost always full of fish bones and scales.

River otters may make extensive journeys across land, even through deep snow. While a river otter looks clumsy on land, it can easily out-run a human with its humped, loping gait. On slippery surfaces, such as wet grass, snow and ice, the otter glides along, usually on its belly with its legs either tucked back or forward to help steer and push. On flat ground, snowslides are sometimes pitted with blurred footprints where the otter has given itself a push for momentum.

In the past, the Northern River Otter's thick, beautiful, durable fur led to exces-

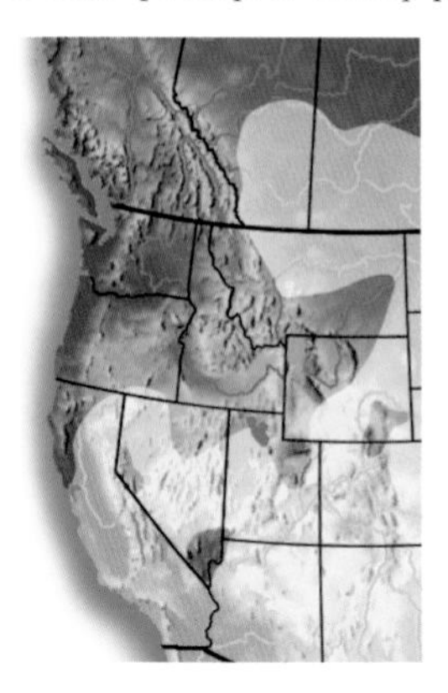

RANGE: The Northern River Otter occurs from near treeline across Alaska and Canada south through forested regions to northern California and northern Utah in the West, and Florida and the Gulf Coast in the East. It is largely absent from the Midwest and Great Plains.

DID YOU KNOW?

When a troupe of agile river otters travel in single file through the water, their undulating, lithe bodies combine to form a very serpent-like form—perhaps with enough similarity to give rise to the rumors of lake-dwelling sea-monsters.

Total Length: 3½–4½ ft (1.1–1.4 m)

Tail Length: 12–20 in (30–51 cm)

Weight: 10–24 lb (4.5–11 kg)

sive trapping that greatly diminished its continental population. Trapping has since been reduced, and the otter seems to be slowly recolonizing parts of North America from which it has been absent for decades. Even in areas where it is known to occur, however, it is infrequently seen, but its marks of playfulness remind us that we are not alone in enjoying the good life.

DESCRIPTION: This large, weasel-like carnivore has dark brown upperparts that look black when wet. It is paler below, and the throat is often silver gray. The head is broad and flattened, and it has small eyes and ears and prominent, whitish whiskers. The feet are webbed. The long tail is thick at the base and gradually tapers to the tip. A male is larger than a female.

HABITAT: Year-round, river otters primarily occur in or along wooded rivers, ponds and lakes, but they sometimes roam far from water. They may be active during the day or night, but they tend to be more nocturnal around human activity. In winter, Northern River Otters almost invariably seek lakes with beaver lodges or bog ponds with steep banks containing old beaver burrows, through which they can enter the water.

FOOD: Crayfish, turtles, frogs and fish form the bulk of the diet, but otters occasionally depredate bird nests and eat small mammals, such as mice, young muskrats and young beavers, and sometimes even insects and earthworms. Otters do not hibernate, and in winter they still chase fish under the surface of the ice.

DEN: The permanent den is often in a bank, with both underwater and abovewater entrances. During its roamings, an otter rests under roots or overhangs, in hollow logs, in the abandoned burrows of other mammals or in abandoned beaver or muskrat lodges. Natal dens are usually in an abandoned muskrat, beaver or Woodchuck dens.

YOUNG: The female bears a litter of one to six blind, fully furred young in March or April. The young are 5 oz (140 g) at birth. They first leave the den at three to four months, and leave their parents at six to seven months. Otters become sexually mature at two years. The mother breeds again soon after her litter is born, but delayed implantation of the embryos puts off the birth until the following spring.

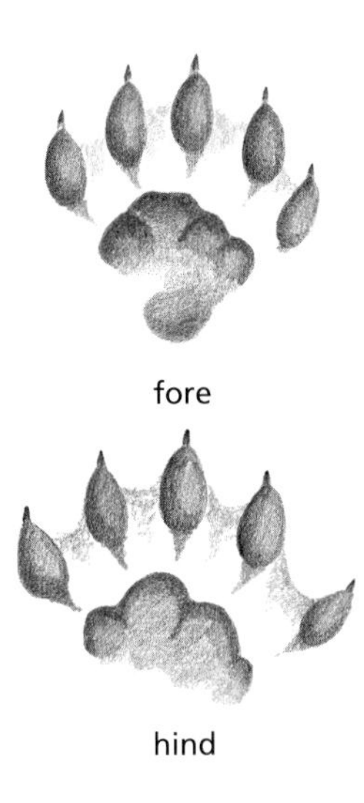

left prints

loping trail

SIMILAR SPECIES: The American Beaver (p. 188) is stouter and has a wide, flat, hairless, scaly tail. The American Mink (p. 102) is smaller, its feet are not webbed, and its tail is cylindrical, not tapered.

American Beaver

Ringtail

Bassariscus astutus

The Ringtail seems to have it all: it is an agile hunter, an all-around athlete and capable of making a meal of almost everything digestible. As well, the striking Ringtail is unlike any other Rocky Mountain mammal in appearance. A mythical description of the Ringtail might say that it has the face of a fox, the body of a weasel, the tail of a lemur, the dexterity of a cat and the eyes of an owl.

The Ringtail is seldom seen by people because it forages almost exclusively at night. It hunts with a style similar to that of a cat. It waits in ambush, and as unsuspecting prey approaches, it readies itself and then pounces on the animal. The Ringtail tries to knock its prey off balance and pin it down to deliver a fatal bite to the neck. This hunting style and the Ringtail's impressive mousing skill are responsible for its alternate common name, "cacomistle." This name is from the language of the Mexican Nahuatl people; it means "half mountain lion." Unlike the Mountain Lion, however, the Ringtail is quite small and hunts mainly rodents, reptiles and amphibians.

Occasionally Ringtails are seen in the headlights of a car, particularly where roads pass over streams and rivers in the southern Rockies. It is also well known for sneaking into camps and backyards and carrying off food items. The Ringtail is remarkably surreptitious, because it seems to equally excel in secrecy as in light-footed nimbleness. A Ringtail uses its semi-retractile, cat-like claws to cling onto almost sheer surfaces, and it can scale trees or nearly vertical cliffs to get at bird nests, reptiles and rodents.

ALSO CALLED: Cacomistle.

DESCRIPTION: Looking like a slender, big-eared, fox-faced cat, the Ringtail is gray or yellowish gray above and buff beneath. Its most noticeable characteristic is the long, bushy tail that is alternately banded black and white. This extremely agile animal almost appears to flow through the rocky terrain it inhabits.

HABITAT: The Ringtail generally occupies rocky slopes, cliffs and canyons in the southern Rockies, usually near water. It has been found as high as 9000 ft (2740 m).

FOOD: The omnivorous Ringtail eats insects and other invertebrates, small mammals, reptiles, amphibians, bird eggs and nestlings, carrion and fruit.

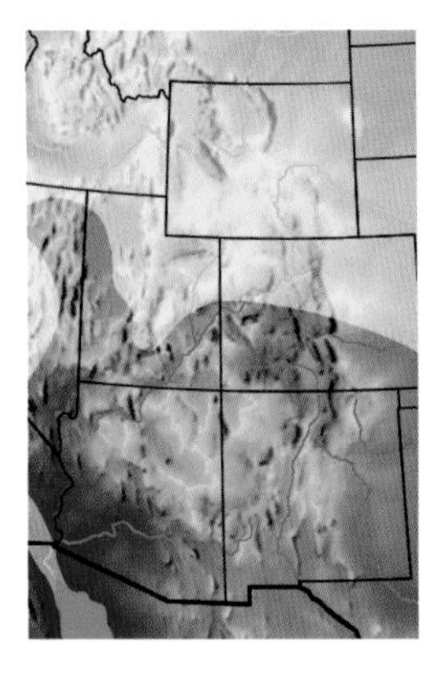

RANGE: Ringtails occur from southern Oregon east through southern Colorado to western Oklahoma and south into southern Mexico.

DID YOU KNOW?

In remote mining operations, miners found the Ringtail's ability to rid the premises of mice so attractive that they put out food to encourage these nocturnal predators. In many instances the Ringtails became so accustomed to sharing the food they were almost tame.

Total Length: 24–32 in (61–81 cm)

Tail Length: 12–17 in (30–43 cm)

Weight: 1¾–2½ lb (0.8–1.1 kg)

walking trail

DEN: The Ringtail's den, which has a tiny entrance, is generally found in rocky debris or in a natural cave, but sometimes a hollow tree or the space beneath an abandoned building is used.

YOUNG: After mating in late February or March, one to five (usually three to four) blind, helpless, 1-oz (28-g) babies are born, usually in May, but sometimes as late as July. Their eyes and ears open and the teeth erupt when they are one month old. At this time, they also switch from a milk diet to animal food brought in by both parents. The mother trains her offspring to hunt, and they disperse when they reach adult size by early winter. Ringtails are sexually mature before their first birthdays.

SIMILAR SPECIES: Only the Common Raccoon (p. 118) has a banded tail like that of the Ringtail, but the raccoon is much stockier and sports a humped back.

left hindprint

Common Raccoon

Procyon lotor

The Common Raccoon is famous for its black bandit mask and ringed tail. The mask suits the raccoon, because it is well known to be a looter of people's gardens, cabins, campsites and, yes, even garbage cans. A raccoon is likely to investigate tasty bits of food and any shiny object it finds. For all its roguish behavior, however, the Common Raccoon has never been associated with ferociousness or savagery—it is mainly a playful and gentle animal unless it is cornered or threatened. Testing a raccoon's ferocity is an unnecessary and simple-minded act, and raccoons have been known to wound and even kill attacking dogs.

Raccoons are among the most frequently encountered wild carnivores in the southern Rocky Mountains. When raccoons are seen, which is usually at night, they quickly bound away, effectively evading flashlight beams and slipping into burrows or climbing to tree retreats. Should their sanctuary be found, raccoons remain still at a safe distance, waiting for the experience to end.

Although personal encounters with the Common Raccoon are less common in the Rocky Mountains than elsewhere, this animal's tendency to frequent muddy environments allows people to find its diagnostic tracks along the edges of wetlands and waterbodies. Like bears and humans, the Common Raccoon walks on its heels, so it leaves unusually large tracks for an animal of its size. It will methodically circumnavigate wetlands in the hopes of finding duck nests or unwary amphibians upon which to dine.

The way that raccoons typically feel their way through the world has long been recognized. In fact, our word "raccoon" is derived from the Algonquin name for this animal, *aroughcoune*, which means "he scratches with his hands." One of the best-known characteristics of the Common Raccoon is its habit of dunking its food in water before eating it. It had long been thought that the raccoon was washing its food—the scientific name *lotor* is Latin for "washer"—but biologists now believe that a raccoon's sense of touch is enhanced by water, and that it is actually feeling for inedible bits to discard.

Long, cold winters are an ecological barrier in the dispersal of this animal, because it does not enter a dormant state in the coldest periods and so requires year-round food availability. Over the past century, however, raccoons have been moving into colder

RANGE: The Common Raccoon occurs from southern Canada south through most of the U.S. and Mexico. It is absent from parts of southern California, central Nevada, Utah and Arizona.

DID YOU KNOW?

Raccoons have thousands of nerve endings in their "hands" and "fingers." It is an asset they constantly put to use, probing under rocks and in crevices for food.

Total Length: 26–38 in (66–97 cm)

Tail Length: 7½–11 in (19–28 cm)

Weight: 12–31 lb (5.4–14 kg)

climes, perhaps due to the increasing human habitation in areas previously ill-suited to them. When raccoons first appeared in Winnipeg, Manitoba, in the 1950s, many people were quite surprised and took them to the local zoo, thinking they were escapees rather than naturally occurring animals.

DESCRIPTION: The coat is blackish- to brownish-gray overall, with lighter, grayish-brown underparts. The bushy tail, with its four to six alternating blackish rings on a yellowish-white background, makes the raccoon one of the most recognizable North American carnivores. There is a black "mask" across the eyes, which is bordered by the white "eyebrows" and mostly white snout, and a strip of white fur separates the upper lip from the nose. The ears are relatively small. Common Raccoons are capable of producing a wide variety of vocalizations: they can purr, growl, snarl, scream, hiss, trill, whinny and whimper.

HABITAT: Raccoons are most often found near streams, lakes and ponds. They are not typically found high in the mountains, because they favor montane woodlands.

FOOD: The Common Raccoon fills the role of medium-sized omnivore in the food web. Besides eating fruits, nuts, berries and insects, it avidly seeks out and eat clams, frogs, fish, eggs, young birds and rodents. Just as a bear does, the raccoon consumes vast amounts of food in fall to build a large fat reserve that will help sustain it over winter.

DEN: The den is often located in a hollow tree, but sites beneath abandoned buildings or under discarded construction materials are increasingly being used. In the foothills, dens can sometimes be found in rock crevices, where grass or leaves carried in by the female may cover the floor.

YOUNG: After about a two-month gestation, the female bears two to seven (typically four) young in late spring. The young weigh just 2 oz (57 g) at birth. Their eyes open at about three weeks, and when they are six to seven weeks old they begin to feed outside the den. At first, the mother carries her young about by the nape of the neck, as a cat carries kittens. About a month later, she starts taking them on extended nightly feeding forays. Some young disperse in fall, but others remain until their mother forces them out when she needs room for her next litter.

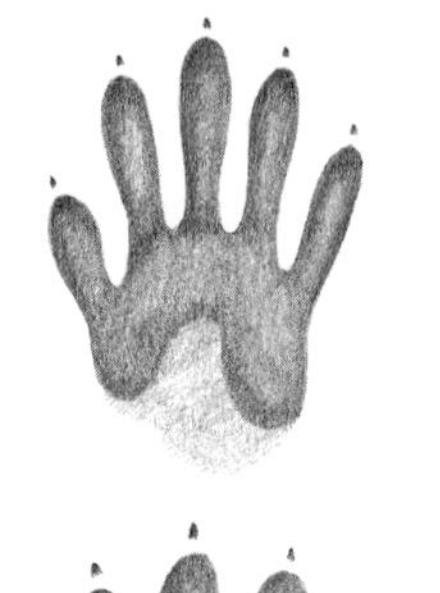

fore

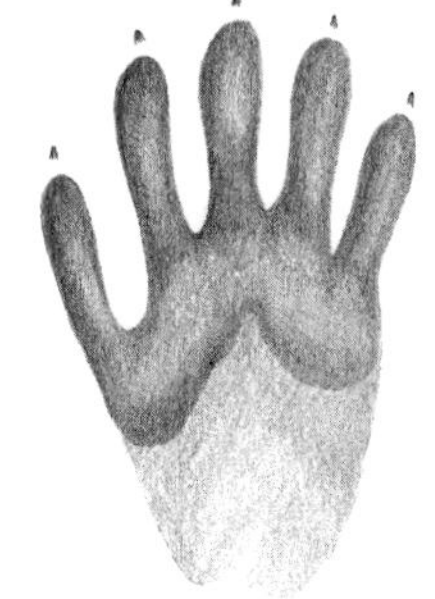

hind

right prints

gallop group

SIMILAR SPECIES: The Common Raccoon is very distinctive. Only the Ringtail (p. 116) could possibly be confused with it, but a Ringtail is much leaner and doesn't have a raccoon's humped back.

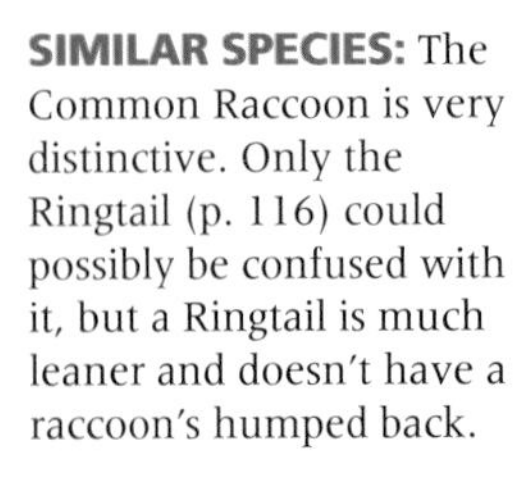

Ringtail

Black Bear

Ursus americanus

The Black Bear, a common inhabitant of forests in the Rocky Mountains, is often feared by city dwellers who come to the mountain parks to appreciate the scenery and wilderness. People who are more experienced with the mountain forest and with the behavior of animals, however, tend to regard the Black Bear with casual familiarity.

Contrary to popular belief and their membership in the carnivore group of mammals, Black Bears do not readily hunt larger animals. They are primarily opportunistic foragers and feed on what is easy and abundant—usually berries, horsetails, other vegetation and insects, although they won't turn up their noses at fish, young fawns or another carnivore's kill. Black Bear sows with young cubs are the most likely to attack young Moose, deer and Elk.

In the past few decades, the infamous dandelion has become increasingly abundant in the mountains along roadsides and swathes cut into the forests, especially in Yellowstone, Banff and Jasper national parks. As a result, Black Bears are now more frequently seen along roadsides, and if a bear looks up to watch your passing car, it will betray to you its new favorite food: with dandelion leaves sticking out of its mouth and the puffy seeds stuck over its face and muzzle, the bear looks like a little kid covered in its favorite ice cream. Unfortunately, together with an increase in bear sightings along roadsides, vehicle collisions that may claim bears' lives are also increasing.

Within its territory, a bear will have favorite feeding places and follow well-traveled paths to these sites. Keep in mind that the trails you hike in the mountains may be used not only by humans, but also by bears wanting to get to lush meadows or rich berry patches.

Normally, Black Bears are peaceful, reclusive animals that will flee to avoid contact with humans if they hear you coming. If you surprise a bear, however, heed its warning of a foot stamp, a throaty "huff" or the champing sound of its teeth. The bear is agitated and probably does not like you, and it is giving you a clear warning to retreat from its territory in respect of its dominance. Most cases of bear attacks (other than those involving a sick bear) occur when these warning signals are not understood by a person who instead remains frozen in place. The bear interprets such behavior as a challenge.

RANGE: Across North America, the Black Bear occurs nearly everywhere there are forests, swamps or shrub thickets. It avoids grasslands and deserts.

DID YOU KNOW?

During its winter slumber, a Black Bear loses 20 to 40 percent of its body weight. To prepare for winter, the bear must eat thousands of calories a day during late summer and fall.

Total Length: 4½–6 ft (1.4–1.8 m)

Shoulder Height: 3–3½ ft (0.9–1.1 m)

Tail Length: 3¼–7 in (8.3–18 cm)

Weight: 88–595 lb (40–270 kg)

One of the darkest threats to Black Bears throughout the world is the aphrodisiac market. Bear paws and gall bladders have high black-market values, and illegal poaching occurs in both Canada and the United States, although to a lesser extent than elsewhere. As the populations of many bears around the world shrink, however, North American bears may be looked on to supply the market. With the price of these alleged libido-lifters already exceeding that of gold, it is feared that North American bear populations, even the generally well-protected animals in the Rocky Mountains, may lose their prestigious stability.

DESCRIPTION: The coat is long and shaggy and ranges from black to brown to honey colored. The body is relatively short and stout. The legs are short and powerful. The large, wide feet have curved, black claws. The head is large and has a straight profile. The eyes are small, and the ears are short, rounded and erect. The tail is very short. An adult male is about 20 percent larger than a female.

HABITAT: Black Bears are primarily forest animals, and their sharp, curved foreclaws enable them to easily climb trees, even as adults. In spring, they often forage in natural clearings or along roadsides.

FOOD: Away from human influences, up to 95 percent of the Black Bear's diet is plant material: leaves, buds, flowers, berries, fruits and roots are all consumed. This bear is like a pig, however, in its ability to consume animal matter, and insects, bees and honey and even young hoofed mammals may be killed and eaten. Carrion and human garbage are eagerly sought out.

DEN: The den, which is only used during winter, may be in a cave or hollow tree, beneath a fallen log or the roots of a wind-thrown tree, or even in a haystack. The bear usually carries in a few mouthfuls of grass to lie on during its sleep. It will not eat, drink, urinate or defecate during its time in the den. The hibernation is not deep; instead, it is like the bear is very groggy or heavily drugged. Rarely, a bear may rouse from this torpor and leave its den on mild winter days.

YOUNG: Black Bears mate in June or July, but the embryos do not implant and begin to develop until the sow enters her den in November. The number of eggs that implant seems to be correlated with the female's weight and condition—fat mothers have more cubs. One to five (usually two or three) young are born and nursed while the sow sleeps. Their eyes open and they become active when they are five to six weeks old. They leave the den with their mother when they weigh 4½–6½ lb (2–3 kg), usually in April. The sow and her cubs generally spend the next winter together in the den, dispersing the following spring. Black Bears typically bear young in alternate years.

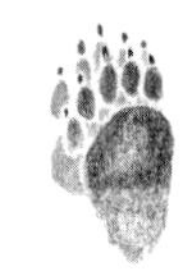

walking trail

SIMILAR SPECIES: The Grizzly Bear (p. 126) is generally larger and has a dished-in face, a noticeable shoulder hump and long, brown to ivory-colored, blunt claws. The Wolverine (p. 104) looks a little like a small Black Bear, but it has a long tail and an arched back.

Grizzly Bear

Grizzly Bear

Ursus arctos

The mighty Grizzly Bear, more than any other animal, makes camping and traveling in the Rocky Mountains an adventure, not just another picnic. Since before the time of Lewis and Clark, Grizzlies have had an uneven lore—a myth and prestige that can be sensed whenever you venture into Grizzly country. Fueled by a mix of fear and curiosity, millions of visitors to mountain parks scan the roadsides and open meadows in the hopes of catching a glimpse of this wilderness icon. Most people leave the parks without a personal grizzly experience, but when a bear is sighted, the human melee that ensues is unlike that which surrounds any other mountain animal. Crowds and "bear jams" are created, further contributing to the aura that surrounds this misunderstood species.

Grizzly Bears are indisputably strong: their massive shoulders and skull anchor muscles that are capable of rolling 200-lb (90-kg) rocks, dragging Elk carcasses and crushing some of the most massive ungulate bones. Ironically, Grizzlies are uncommonly obliged to feed in this manner. Instead, their routine and docile foragings are concentrated on roots, berries and grasses. An adaptable diner, the Grizzly changes its diet from spring through fall to match the availability of foods. For instance, it eats huge quantities of berries when they are available in late summer. A bear swallows many of the berries whole, and its scat often ends up looking like blueberry pie filling. During this time of feasting, a Grizzly's weight may increase by 2 lb (0.9 kg) a day, preparing it for the long winter ahead. It will remain active through fall, until the bitter cold of November limits food and favors sleep.

Although the mountain parks boast the highest numbers of bears, the status of Grizzly Bears in the contiguous states is uncertain. No one can predict their future, but we can increase their chances of survival. Some of the seminal work on Grizzlies dates back to the foundation of conservation biology. In working with these large carnivores, pioneering biologists invented tagging, radio-tracking and other research techniques that have benefited not only the Grizzly, but many other carnivores, including the Bengal Tiger and the Polar Bear.

ALSO CALLED: Brown Bear.

DESCRIPTION: The usually brownish to yellowish coat typically has white-

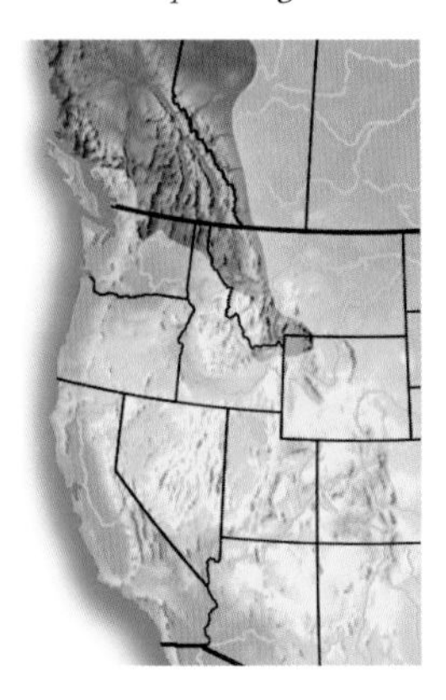

RANGE: In North America, this bear is largely confined to Alaska and northwestern Canada, with montane populations extending south into extreme northern Washington, Idaho Montana and Wyoming. It formerly ranged much further, but Euroamerican settlers extirpated the open-area populations.

DID YOU KNOW?

Because an adult Grizzly's long foreclaws are typically blunt from digging, it cannot easily climb trees. If you think you can escape a bear by climbing a tree, however, you better climb high, because some Grizzlies can reach 12 ft (3.6 m) up a tree trunk.

Total Length: 6–8½ ft (1.8–2.6 m)

Shoulder Height: 3–4 ft (0.9–1.2 m)

Tail Length: 3–7 in (7.6–18 cm)

Weight: 242–1160 lb (110–530 kg)

tipped guard hairs that give it a grizzled appearance (from which the name "grizzly" is derived). Some individuals are completely black; others can be nearly white. The face has a concave profile. The eyes are relatively small and the ears are short and rounded. A large hump at the shoulder makes the forequarters higher than the rump in profile. The large, flat paws have long claws; at least some of the front claws can be more than 2 in (5 cm) long.

HABITAT: Originally, most Grizzly Bears were animals of the prairies and open areas, where they used their long claws to dig up roots, bulbs and the occasional burrowing mammal. Although their current range is largely forested, mountain bears often forage on open slopes and in the alpine tundra.

FOOD: Although 70 to 80 percent of a Grizzly's diet is plants, including leaves, stems, flowers, roots and fruits, it eats more animals, including other mammals, fish and insects, than its black cousin does. A Grizzly may dig insects, ground squirrels, marmots and even mice out of the ground. Young hoofed mammals are eagerly sought by sow bears with cubs, and even large adult cervids and Bighorn Sheep may be attacked and killed. Particularly after it emerges from its winter sleep, a bear is attracted to carrion, which it can smell from 10 mi (16 km) away. The Grizzly Bear can eat huge meals of meat: one adult consumed an entire road-killed Elk in four days.

walking trail

DEN: Most dens are on north- or northeast-facing slopes, in areas where snowmelt does not begin until late April or early May. The den is usually in a cave, or it is dug into tree roots. The bear enters its den in late October or November, during a heavy snowfall that will cover its tracks. The bear soon falls asleep; it will not eat, drink, urinate or defecate for six months.

YOUNG: A sow has litters in alternate years, typically having her first after her seventh birthday. Grizzlies mate in June or July, but with delayed implantation of the embryo, the cubs are not born until some time between January and early March, when the mother is asleep in her den. The one to four (generally two) cubs are born naked, blind and helpless. They nurse and grow while their mother continues to sleep, and they are ready to follow her when she leaves the den in April or May. A sow and her cubs typically den together the following winter.

SIMILAR SPECIES: The Black Bear (p. 122) is generally smaller, is tallest at the rump (it doesn't have a humped shoulder) and has a straight profile and shorter, black, curved claws.

Black Bear

Coyote

Canis latrans

A chorus of yaps, whines, barks and howls complements the darkening skies over much of the Rocky Mountains. Although Coyote calls are most intense during late winter and spring, corresponding to courtship, these maniacal sounds can be heard during suitable weather at any time of the day or year. Often initiated by one animal, many family groups soon join in, and the calls pour from the valleys, making it obvious to all that these animals relish getting together and making noise. In these vocalizations, you can hear the abundance of Coyotes and their presence in the landscape.

When Lewis and Clark and David Thompson traveled through the Rocky Mountain wilderness two centuries ago, they made frequent references in their journal entries to foxes and wolves, but they seldom mentioned Coyotes. Coyotes have increased their numbers across North America in the past century in response to the expansion of agriculture and forestry and the reduction of wolf populations. Despite widespread human efforts to exterminate them, they have thrived.

One of the few natural checks on Coyote abundance seems to be the Gray Wolf. As the much larger and more powerful canids of the wilderness neighborhood, wolves typically exclude Coyotes from their territories. Prior to the 19th century, the natural condition favored wolves, but changes in the region since then have greatly benefited Coyotes. They are now so widely distributed and comfortable with human development that almost every mountain valley holds a healthy population.

Because of its relatively small size, the Coyote typically preys on small animals, such as mice, voles, ground squirrels, birds and hares, but it has also been known to kill Bighorn Sheep and deer, particularly their young. Although they usually hunt alone, Coyotes occasionally form packs, especially when they hunt hoofed mammals during winter. The Coyotes may split up, with some waiting in ambush while the others chase the prey toward them, or they may run in relays to tire their quarry—the Coyote, which is the best runner among the North American canids, typically cruises at 25–30 mph (40–50 km/h).

Coyotes owe their modern-day success to their varied diet, early age of first breeding, high reproductive output and flexible living requirements. They consume carrion throughout the year, but they also feed on such diverse offerings

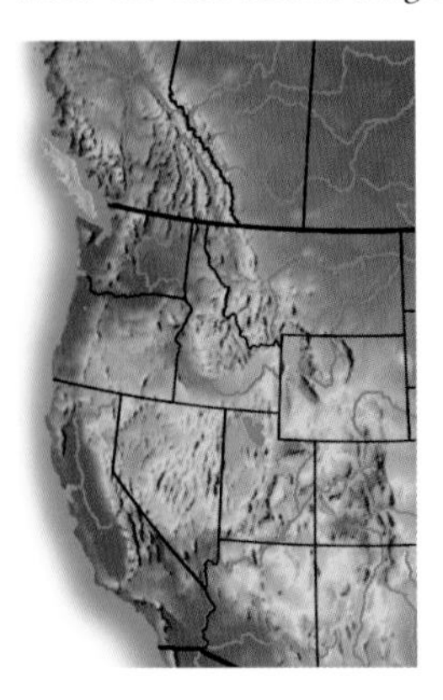

RANGE: Coyotes are not found in the western third of Alaska, the tundra regions of northern Canada or the extreme southeastern United States; their range essentially covers the remainder of North America.

DID YOU KNOW?

True to its adaptive nature, the Coyote has been known to form hunting parties, in wolf fashion, to hunt down large animals that would be too powerful for a lone individual to subdue.

Total Length: 3½–4½ ft (1.1–1.4 m)
Shoulder Height: 23–26 in (58–66 cm)
Tail Length: 12–16 in (30–41 cm)
Weight: 18–44 lb (8.2–20 kg)

as eggs, mammals, birds and berries. Their variable diet and nonspecific habitat choices allow them to adapt to just about any region of North America.

Coyotes can, and do, interbreed with domestic dogs. The "coydog" offspring often become nuisance animals, killing domestic livestock and poultry.

DESCRIPTION: Coyotes look like gray, buffy or reddish-gray, medium-sized dogs. The nose is pointed and there is usually a gray patch between the eyes that contrasts with the tawny top of the snout. The bushy tail has a black tip. The underparts are light to whitish. When frightened, a Coyote runs with its tail tucked between its hindlegs. Coyotes in the northern part of the Rockies tend to be considerably larger than those in southern populations.

HABITAT: Coyotes are found in all terrestrial habitats in North America except the barren tundra of the far north and the humid southeastern forests. They have greatly expanded their range in part where humans extirpated the Gray Wolf and in part because the clearing of forests brought about changes in habitat that made it easier for Coyotes to establish populations in new areas.

FOOD: Although primarily carnivorous, feeding on squirrels, mice, hares, birds, amphibians and reptiles, Coyotes will sometimes eat cactus fruits, melons, berries and vegetation. Most ranchers dislike coyotes because they frequently take sheep, calves and pigs that are left exposed. They may even attack and consume dogs.

DEN: The den is usually a burrow in a slope, frequently an American Badger or Woodchuck hole that has been expanded to 1 ft (30 cm) in diameter and about 10 ft (3 m) deep. Rarely, Coyotes have been known to den in an abandoned car, a hollow tree trunk or a dense brush pile.

YOUNG: A litter of 3 to 10 (usually 5 to 7) pups is born between late March and late May, after a gestation of about two months. The furry pups are blind at birth. Their eyes open after about 10 days, and they leave the den for the first time when they are three weeks old. Young Coyotes fight with each other and establish dominance and social position at just three to four weeks of age.

foreprint

gallop group

SIMILAR SPECIES:
The Gray Wolf (p. 134) is generally larger, has much bigger feet and longer legs and carries its tail straight back when it runs. The Red Fox (p. 138) is generally smaller and has a white tail tip and black forelegs. Coyote-like domestic dog breeds generally have more bulging foreheads and usually carry their tails straight back when they run.

Red Fox

Gray Wolf

Canis lupus

For many North Americans, the Gray Wolf represents the apex of wilderness, symbolizing the pure, yet hostile, qualities of all that remains wild. Other people disparage this iconic representation, characterizing wolves as blood-lusting enemies of domestic animals and the ranchers who care for them. Objective opinions about the Gray Wolf are few; caricatures, whether positive or negative, abound. The truth probably lies in the words of Aldo Leopold: "Only the mountain has lived long enough to listen objectively to the howl of the wolf."

Regardless of its true nature, the Gray Wolf has unknowingly been the subject of some of the most progressive and politically controversial conservation programs in North America. The most noteworthy efforts have been in Yellowstone National Park, where wolves taken from Canada were reintroduced into the park. This effort has many political, ecological and social ramifications, and although the park's ecosystem has benefited, the wolves themselves may not. Had wolves returned to the park through natural dispersal from the north, they would have had the full protection of the Endangered Species Act. As a reintroduced species, however, there are legal and jurisdictional problems that limit the protection of Yellowstone's "experimental" wolf population. By January 2000, there were a few more than 100 wolves in Yellowstone.

One practical benefit of reintroducing wolves into Yellowstone is to restore predation to the hoofed mammal populations. Part of this effect relates to a wolf pack's tendency to specialize in killing a single species. The wolves transplanted into Yellowstone have been killing mostly Elk, even though American Bison abound. Wolves, like a small assortment of other predators and some raptorial birds, seem to develop "search images" that focus them on one or two prey species. In both Alaska and the Yukon, wolves have been shown to reduce prey populations of Moose and Caribou to about 20 percent of the habitat's carrying capacity. When 70 percent of the wolves were killed in these areas, prey populations rebounded to about the carrying capacity of the region.

Some people have suggested that human society has a near equal in the social structure of the wolf pack. A pack behaves like a "super organism," cooperatively making it possible for more

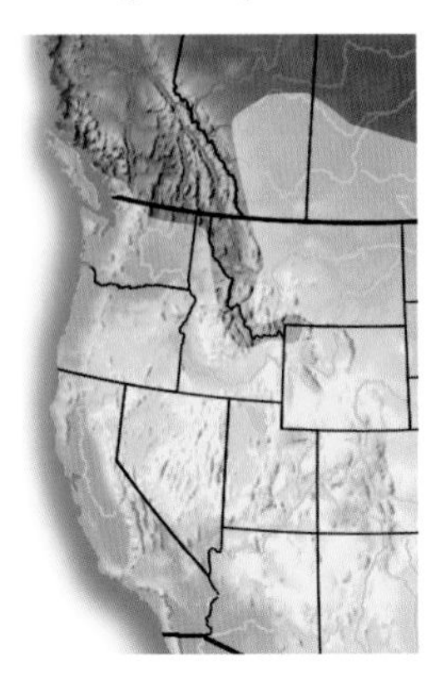

RANGE: Much reduced from historic times, the Gray Wolf's range currently covers most of Canada and Alaska, except on the prairies and in the southern parts of eastern Canada. It extends south into Minnesota and Wisconsin, and along the Rocky Mountains into Idaho, Montana and Wyoming.

DID YOU KNOW?

Wolves are capable of many facial expressions, such as pursed lips, smile-like, submissive grins, upturned muzzles, wrinkled foreheads and angry, squinting eyes. Wolves even stick their tongues out at each other as a gesture of appeasement or submission.

Total Length: 4½–6½ ft (1.4–2 m)
Shoulder Height: 26–38 in (66–97 cm)
Tail Length: 14–20 in (36–51 cm)
Weight: 57–170 lb (26–77 kg)

wolves to survive in an area. By hunting together, pack members can catch and subdue much larger prey than could one wolf acting alone. A social hierarchy is strictly followed in the pack, and moments of tension rarely break the orderly, communal pack scene. Often, only the top animals (called the "alpha" male and female by wolf biologists) will reproduce, but other pack members help with bringing food to the pups and defending the group's territory against the intrusions of other wolves.

A wolf pack generally occupies a large territory—usually 100–300 mi^2 (260–780 km^2)—so individual densities are extremely low. Howling is an effective way for wolves to keep in contact over long distances, and it appears to play an important role in communication among pack members and between adjacent packs. In the sound of a wolf's long, drawn-out howl is the spirit of the untamable and the independence of all wild animals. When you are next in wild country, cup your hands and offer your best howl to the night sky. With any luck, a wolf may reply, and, in that electric voice, whatever preconception you brought to the encounter will be forever changed.

ALSO CALLED: Timber Wolf.

DESCRIPTION: Most wolves resemble a long-legged German Shepherd dog with extra-large paws. Although typically thought of as being a grizzled gray color, a wolf's coat can range from coal black to creamy white. Black wolves are most common in dense forests; whitish wolves are characteristic of the high Arctic. The bushy tail is carried straight behind the wolf when it runs. In social situations, the height of the tail generally relates to the social status of that individual.

HABITAT: Although wolves formerly occupied grasslands, forests, deserts and tundra, they are now mostly restricted to forests, streamside woodlands and arctic tundra.

FOOD: Because large carnivores customarily eat large herbivores, wolves are destined to eat cervids and Bighorn Sheep. Although large mammals typically comprise about 80 percent of the diet, wolves also devour rabbits, mice, nesting birds and carrion when it is available. Where humans leave livestock unguarded, wolves may take cattle, sheep, goats, horses, dogs, cats and pigs.

DEN: Wolf dens are usually located on a rise of land near water. Most dens are bank burrows, and they are often made by enlarging the den of a fox or burrowing mammal. Sometimes a rock slide, hollow log or natural cave is used. Sand or soil scratched out of the entrance by the female is usually evident as a large mound. The burrow opening is generally about 23 in (58 cm) across, and the burrow extends back 6½–33 ft (2–10 m) to a dry natal chamber with a floor of packed soil. The beds from which adults can keep watch are generally found above the entrance.

YOUNG: A litter generally contains five to seven pups (with extremes of 3 to 13), which may be of different colors. The newborn pups resemble domestic dogs in their development: their eyes open at 9 to 10 days, and they are weaned at six to eight weeks. The pups are fed regurgitated food until they begin to follow the pack on hunts. Wolves become sexually mature a couple of months before their third birthdays, but the pack hierarchy largely determines their first incidence of mating.

foreprint

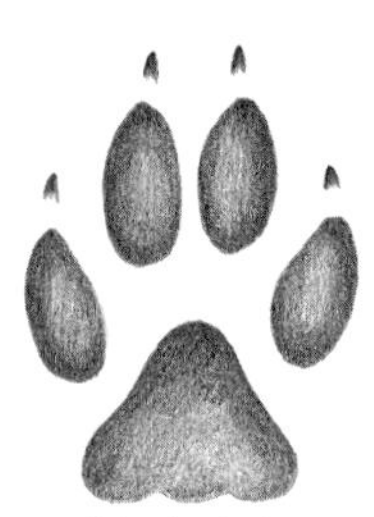
hindprint

SIMILAR SPECIES:
The Coyote (p. 130) is smaller and has a reddish patch on the nose, a feature not seen in wolves. Wolf tracks are 4 in (10 cm) or more from front to back; Coyote tracks are never more than 3 in (7.6 cm) long.

Coyote

Red Fox

Vulpes vulpes

More than the Rocky Mountains' other native canids, the Red Fox has received some favorable presentations in literature and modern culture. From Aesop's Fables to construction yard compliments, the fox is often symbolized as a cunning, intelligent, attractive and noble animal. Foxes have a well-deserved nickname, "reynard," from the French word *renard*, which refers to someone who is unconquerable owing to their cleverness. The fox's intelligence, undeniable comeliness and positive impact upon most farmlands have endeared it to many people who otherwise may not notice wildlife.

To watch a fox in its natural habitat is an experience that embodies playfulness, roguishness, stealth and drama. Young fox kits at their den wrestle and squabble in determined sibling rivalry. If its siblings are busy elsewhere, a young kit may amuse itself by challenging a plaything, such as a stick or piece of old bone, to a bout of aggressive mock combat. An adult out mousing will sneak up on a rustling mouse in the grass and jump stiff-legged into the air, hoping to come down directly atop the unsuspecting rodent. If the fox misses, it stomps and flattens the grass with its forepaws, biting in the air to try to catch the mouse. Usually the fox wins, but, if not, it will slip away with stately composure as though the display of undignified abandon never occurred.

Oddly enough, Red Foxes exhibit both feline dexterity and a feline hunting style. Foxes may hunt using an ambush style, or they may creep along in a crouched position, ready to pounce on unsuspecting prey. Another un-dog-like characteristic is the large gland above the base of the tail, which gives off a strong musk somewhat resembling the smell of a skunk. This scent is responsible for fox hounds being able to easily follow a fox. Foxes are territorial, and the males, like other members of the dog family, mark their territorial boundaries with urine.

Despite its vast range, the Red Fox is rarely seen. Its primarily nocturnal activity is probably the main reason, but a fox's keen senses of sight, hearing and smell enhance its elusive nature. Winter may be the best time to see a fox: it is more likely to be active during the day, and its color stands out when it is mousing in a snow-covered field.

Red Foxes have adapted to human activity, and most of them live in farming communities and even in cities. A few Red Foxes, however, enter the

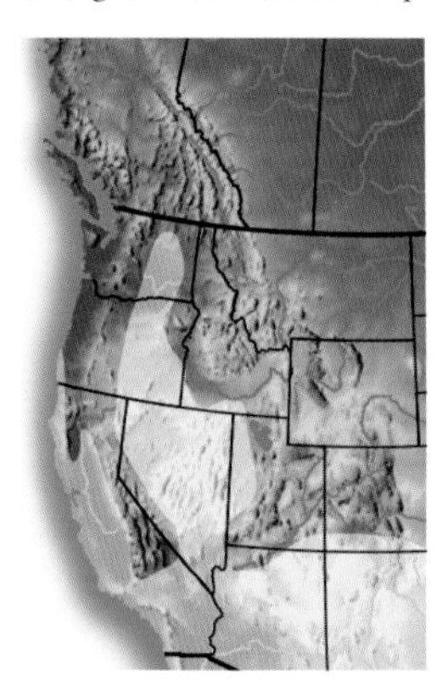

RANGE: In North America, this holarctic species occurs throughout most of Canada and the U.S., except for the high Arctic, northwestern British Columbia and much of the western U.S.

DID YOU KNOW?

The Red Fox's signature feature—its white-tipped, bushy tail—provides balance when the fox is running or jumping, and during cold weather a fox wraps its tail over its face.

Total Length: 35–44 in (89–112 cm)

Shoulder Height: 15–16 in (38–41 cm)

Tail Length: 14–17 in (36–43 cm)

Weight: 8–15 lb (3.6–6.8 kg)

wilderness areas of the Rocky Mountains, where these diminutive carnivores live on mice and carcasses in the shadow of the Gray Wolf.

DESCRIPTION: This small, slender, dog-like fox has an exceptionally bushy, long tail. Its upperparts are usually a vivid reddish orange, with a white chest and belly, but there are many color variations: there is a Coyote-colored phase in parts of the Rocky Mountains; the "cross fox" has darker hairs along the back and across the shoulder blades; and the "silver fox," which is more frequently encountered on the prairies, is mostly black with silver-tipped hairs. In all color phases, the tail has a white tip and the backs of the ears and fronts of the forelegs are black.

HABITAT: The Red Fox prefers open habitats interspersed with brushy shelter year-round. It avoids extensive areas of dense, coniferous forests with heavy snowfall.

FOOD: This opportunistic feeder usually stalks its prey and then pounces on it or captures it after a short rush. In winter, small rodents, rabbits and birds make up most of the diet, but dried berries are also eaten. In more moderate seasons, invertebrates, birds, eggs, fruits and berries supplement the basic small-mammal diet.

DEN: The Red Fox generally dens in a burrow, which the vixen either digs herself or, more usually, makes by expanding a marmot or badger hole.

The den is sometimes located in a hollow log, in a brush pile or beneath an unoccupied building.

YOUNG: A litter of 1 to 10 kits is born in April or May after a gestation of about 7½ weeks. The kits weigh about 3½ oz (99 g) at birth. Their eyes open after nine days, and they are weaned when they are one month old. The parents first bring the kits dead food and later crippled animals. The father may bring back to the den several voles, or perhaps a hare and some mice, at the end of a single hunting trip. After the kits learn to kill, the parents start taking them on hunts. The young disperse when they are three to four months old; they become sexually mature well before their first birthdays.

foreprint

side-trotting trail

SIMILAR SPECIES: The larger Coyote (p. 130), the similarly-sized Common Gray Fox (p. 146) and the smaller Swift Fox (p. 142) all have dark-tipped tails and do not have black forelegs.

Common Gray Fox

Swift Fox

Vulpes velox

Encountering a Swift Fox in the wild is an unlikely event, mainly because these cat-sized canids are nocturnal. If they are active at all during the day, it is to sun themselves outside their dens. They rest and bask in a curled-up position, much like a cat, but they are always alert and never far from the safety of their dens. Several family members occupy the same den, and large dens may house more than one family. Some populations of Swift Foxes seem almost colonial, with many families living and feeding together.

The Swift Fox is one of several mammals that suffered extensive population and range reductions during the time of agricultural expansion in North America. Historically, the Swift Fox was found throughout the plains and sparsely wooded regions, ranging as far north as the Red Deer River in Alberta. Very little is known about the current distribution and population sizes of the Swift Fox, but it is found in the Rocky Mountains south from Wyoming.

Although this fox is not highly prized for its coat, its numbers declined from habitat loss as prairies were converted to agricultural fields and because of deaths from traps and poison set for Gray Wolves and Coyotes. Probably fewer Swift Foxes were ever killed than Coyotes, but Swift Foxes never fully recovered their original numbers. Part of this disparity is related to the short life expectancy for Swift Foxes, but it is also related to their mating strategy. Swift Foxes live as mated pairs, and if its mate is killed, the remaining partner may not mate again. Coyotes have a much different mating strategy. They live either alone or in small, hierarchical groups. The death of one or more Coyotes stimulates the others to mate, so the population remains stable or even increases.

The Swift Fox has high mortality rates, especially from larger grassland predators—ignoring its family ties, the Coyote is the Swift Fox's main predator and accounts for the majority of known deaths. To avoid Coyote predation, the Swift Fox tends to den in areas with sweeping vistas, which give the foxes ample visual warning of approaching danger. These innate adaptations, coupled with a progressive implementation of effective conservation strategies, will hopefully ensure the future of this unique fox.

Many Swift Fox reintroduction programs have recently had some success with increasing their numbers in the wild. Many of the first Swift Foxes

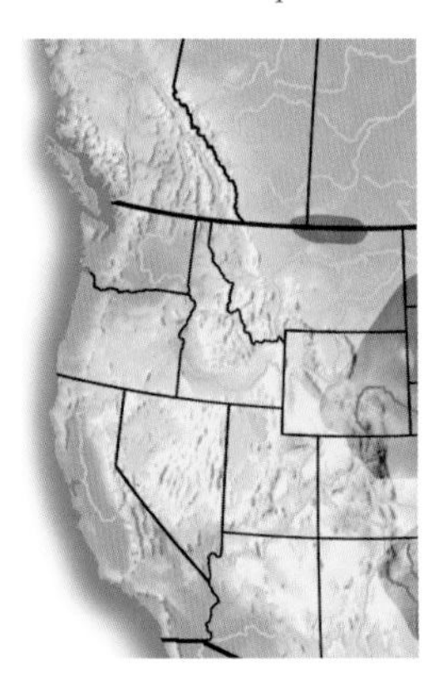

RANGE: This fox's historic range included the arid regions of the prairie provinces south to northern Texas. Although its numbers are greatly reduced, the Swift Fox may still be found scattered sparsely throughout this range.

DID YOU KNOW?

Because Swift Foxes mate for life, they are extremely fastidious about mate selection. The initial courtship of a pair can last for up to three months before the couple decides they are right for each other and mate.

Total Length: 28–36 in (71–91 cm)

Shoulder Height: 11–12 in (28–30 cm)

Tail Length: 10–14 in (25–36 cm)

Weight: 4–6 lb (1.8–2.7 kg)

raised in captivity and released into the wild rarely survived long, but relocated, wild-caught individuals have been more successful in the reintroduction programs. Unfortunately, reestablished foxes may be so isolated that there is little chance of population growth. The reintroduction of Swift Foxes into areas of continuous, suitable habitat would likely be necessary for this animal to regain its pre-settlement numbers. Such a situation is unlikely, however, because there are insufficiently extensive areas of Swift Fox habitat remaining.

DESCRIPTION: The summer coat of this tiny fox closely matches its environment: the back is mainly pale rufous or buffy gray, and the sides are a lighter yellowish buff. The long guard hairs are tipped with either white or black, giving the fox an overall grizzled appearance. Each side of the muzzle bears a distinct black spot, and the neck, backs of the ears and legs have orange highlights. The bushy, long tail has a noticeable black tip.

HABITAT: This open-country fox inhabits prairies, badlands and other arid areas. In the Rockies, Swift Foxes may occur in open areas of the foothills region, but they are never found in dense montane forests.

FOOD: Rabbits and small rodents make up the bulk of the diet, but Swift Foxes also eat birds, insects and even grass or berries. They seem to be quite adept at catching grouse and prairie-chickens in snowy conditions. Other food items include small lizards and amphibians.

The largest animals that Swift Foxes can prey on are jackrabbits.

DEN: A Swift Fox's fairly complex burrow usually has four or five entrances. The same burrow is used throughout the year, and perhaps even for a lifetime. Some burrows become very large and can house more than one family. The burrow is also home to an assortment of other animals, such as invertebrates and even toads. Abandoned burrows become homes for skunks, snakes, mice and Burrowing Owls.

YOUNG: The kits are born in a small, bare chamber of the burrow after a gestation of about seven to eight weeks. The litter of three to six young requires great care for the first few weeks. The kits open their eyes and ears at two weeks, and they are weaned at six weeks. They disperse in fall and may breed before they are one year old.

side-trotting trail

SIMILAR SPECIES: The Red Fox (p. 138) is larger and has black forelegs and a white-tipped tail. The Common Gray Fox (p. 146) is larger and has reddish undersides. The Coyote (p. 130) is much larger and has a tawny patch on the upper surface of its snout.

Red Fox

Common Gray Fox

Urocyon cinereoargenteus

Truly a crafty canid, the Common Gray Fox is known to elude predators by taking the most unexpected of turns—running up a tree. Unlike other canids, this fox seems comfortable in a tree, and it may climb into the branches to rest and sleep. There are even rare records of these foxes denning in natural tree cavities and raising their litters as high as 20 ft (6.1 m) off the ground. It is the only North American member of the dog family that can climb.

The Common Gray Fox's tendency to live in areas of tree cover means it is less frequently seen than other foxes. Furthermore, the majority of its activity occurs during the dark or twilight hours, enhancing its clandestine nature. In the Rocky Mountains, this handsome fox is found in only a handful of locations. A visit to the mountains of Colorado, New Mexico or even Utah might prove fruitful for seeing one. Remember, you may have to turn your eyes skyward and scan the trees, especially those with thick, heavily forked trunks or leaning branches.

After the mating season, the male stays with the female and helps raise the young. His primary role after the female gives birth is to bring her food, because she must remain with the young constantly for several days. Gray foxes often cache their food, especially large kills that cannot be consumed at once. Small kills may be buried right near the den during the whelping season, partly to provide the female with ready food but also to stimulate the interest of the pups. Large cache sites are made either in heaped up vegetation or in holes dug into loose dirt.

Many of the fox populations in North America suffered great losses during the peak of the fur trade. To most foxes, Coyotes and wolves, humans are the worst enemy. Fortunately for the Common Gray Fox, its pelt is of lower quality (to humans, that is—certainly the fox appreciates it). Its beautifully patterned, grizzled fur is very stiff, rather than soft and long like the winter coat of a Red Fox, so it was trapped less. The Common Gray Fox has also suffered much less persecution from farmers, because it is quite shy and rarely hunts domestic animals. Unlike the Red Fox, the Common Gray Fox is not inclined toward chickens. It prefers to hunt the mice that abound around a henhouse, and its mousing ability is so good that it is even considered a welcome visitor to a barnyard.

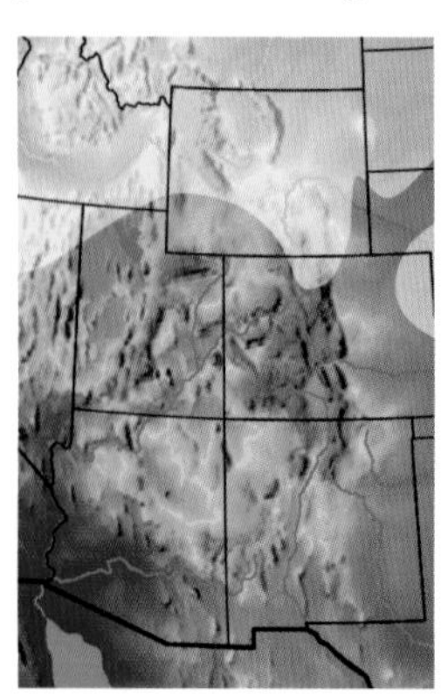

RANGE: Common Gray Foxes have an extensive distribution in the U.S. They range through most of the eastern states, from Texas west to California and up the West Coast to Oregon. Some populations can be found in Colorado and Utah.

DID YOU KNOW?

Although this fox is quite small, it can run very quickly over short distances. In one record, a Common Gray Fox topped 28 mph (45 km/h).

Total Length: 31–44 in (79–112 cm)

Shoulder Height: 14–15 in (36–38 cm)

Tail Length: 11–17 in (28–43 cm)

Weight: 7½–13 lb (3.4–5.9 kg)

DESCRIPTION: This handsome fox has an overall grizzled appearance because of the long, grayish fur over its back. Its undersides are reddish in color, as are the back of the head, throat, legs and feet. Sometimes the belly may be mostly white with only reddish highlights. The tail is gray or even black on top, with reddish undersides and a black tip. The ears are pointed and mainly gray in color, with patches of red on the back side. A distinct black spot is present on either side of the muzzle.

HABITAT: The Common Gray Fox inhabits a variety of different environments, always near trees or ground cover. This fox prefers foraging in wooded areas rather than open environments.

DEN: This fox's den can be found in a variety of places. Most commonly, it is on a ridge or rocky slope or under brush cover in a woodland, but it may also be underground. If necessary, a fox digs a burrow itself, but more often it refurbishes the abandoned burrow of another animal, such as a Woodchuck or American Badger.

FOOD: Gray foxes are more omnivorous than other foxes. They consume a variety of small mammals, such as rabbits, rodents and birds, as well as large amounts of insects and other invertebrates. Late in summer, grasshoppers, crickets and other agricultural pests constitute much of the diet. A significant part of the diet is vegetable matter, such as fruits and grasses. Favorite items include apples, grapes, persimmons and nuts.

BREEDING: Mating occurs in January or February, and after a gestation of about 53 days, one to seven young are born. They are born blind and almost hairless, and for the first several days they require constant care by the mother. After 12 days their eyes open, and they venture out of the den when they are 1½ or 2 months old. When they are four months old, they learn how to hunt and accompany their parents while foraging. By the fifth month, they have dispersed to start their own dens.

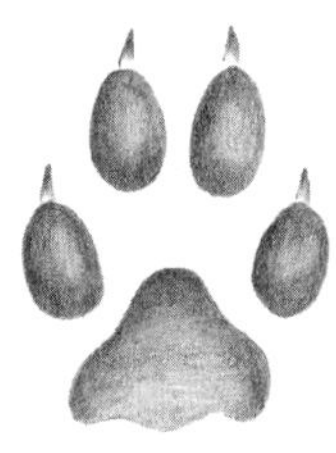

foreprint

hindprint

trotting trail

SIMILAR SPECIES: The Swift Fox (p. 142) is smaller and lacks the reddish undersides. The Red Fox (p. 138) is much redder overall and has a white-tipped tail. The Coyote (p. 130) is larger and lacks the black spots on either side of the muzzle.

Swift Fox

RODENTS

In terms of sheer numbers, rodents are the most successful group of mammals in the Rockies. Because we usually associate rodents with rats and mice, the group's most notorious members, many people look on all rodents as filthy vermin. You must remember, however, that the much more endearing chipmunks, marmots, beavers and squirrels are also rodents.

A rodent's best-known features are its upper and lower pairs of protruding incisor teeth, which continue to grow throughout the animal's life. These four teeth have pale yellow to burnt orange enamel only on their front surfaces; the soft dentine at the rear of each tooth is worn away by the action of gnawing, whereby the teeth retain knife-sharp cutting edges along the forward edge. Most rodents are relatively small mammals, but beavers and porcupines can grow quite large.

Common Porcupine

Porcupine Family
(Erethizontidae)

The stocky-bodied Common Porcupine has some of its hairs modified into sharp-pointed quills that it uses in defense. Its sharp, curved claws and the rough soles of its feet are adapted for climbing.

Western Jumping Mouse

Jumping Mouse Family
(Zapodidae)

Jumping mice are so called because they make long leaps when they are startled. Their hindlegs are much longer than their forelegs, and the tail, which is longer than the combined length of the head and body, serves as a counterbalance during jumps. Jumping mice are almost completely nocturnal.

Western Harvest Mouse

Mouse Family
(Muridae)

This diverse group of rodents is the largest and most successful mammal family in the world. Its members include the familiar rats and mice, as well as voles and lemmings. The Rocky Mountain representatives of this family vary in size from the tiny Western Harvest Mouse to the Common Muskrat.

Beaver Family
(Castoridae)

The American Beaver accounts for half the worldwide species in this family. It is the largest North American rodent, and it is one of the most visible mammals in the region. After humans, it is probably the animal with the biggest impact on the landscape of many states and provinces.

American Beaver

Pocket Mouse Family
(Heteromyidae)

Pocket mice and kangaroo rats make up a group of small to medium-sized rodents that are somewhat adapted to a subterranean existence. They feed mainly on seeds, and they use their fur-lined cheek pouches to transport food to caches in their burrows. Typically occurring in dry environments, many of them can live for a long time without drinking water.

Olive-backed Pocket Mouse

Pocket Gopher Family
(Geomyidae)

Almost exclusively subterranean, all pocket gophers have small eyes, tiny ears, heavy claws, short, strong forelegs and a short, sparsely haired tail. Their external, fur-lined cheek pouches, or "pockets," are primarily used to transport food. The lower jaw is massive, and the incisor teeth are used in excavating tunnels.

Northern Pocket Gopher

Squirrel Family
(Sciuridae)

This family, which includes chipmunks, tree squirrels, flying squirrels, marmots and ground squirrels, is considered the most structurally primitive group of rodents. All its members, except the flying squirrels, are active during the day, so they are seen more frequently than other rodents.

Red-tailed Chipmunk

Common Porcupine

Erethizon dorsatum

Although it is aggressiveness that is most often celebrated among Rocky Mountain mammals, the trademark of the Common Porcupine is its unsurpassed defensive mechanism. A porcupine's formidable quills, numbering about 30,000, are actually modified, stiff hairs with overlapping, shingle-like barbs at their tips. Contrary to popular belief, a porcupine cannot throw its quills, but if it is attacked, it will lower its head in a defensive posture and lash out with its tail. The loosely rooted quills detach easily, and they may be driven deeply into the attacker's flesh. The barbs swell and expand with blood, making the quills even harder to extract. Quill wounds may fester, or, depending on where the quills strike, they can blind an animal, prevent it from eating or even puncture a vital organ.

Porcupines are strictly vegetarian, and they are frequently found feeding in agricultural fields, willow-edged wetlands and forests. The tender bark of young branches seems to be a porcupine delicacy, and although you wouldn't think it from their size, porcupines can move far out on very thin branches with their deliberate climbing. Accomplished, if slow, climbers, porcupines use their sharp, curved claws, the thick, bumpy soles of their feet, and the quills on the underside of the tail in climbing. These large, stocky rodents often remain in individual trees and bushes for several days at a time, and when they leave a foraging site, the naked cream-colored branches are clear evidence of their activity.

The Common Porcupine is mostly nocturnal, and it often rests by day in a hollow tree or log, in a burrow or in a treetop. It is not unusual to see a porcupine abroad by day, however, either in an open field or in a forest. It often chews bones or fallen antlers for calcium, and the sound of a porcupine's gnawing can sometimes be heard at a considerable distance.

Unfortunately for the Common Porcupine, its armament is no defense against vehicles—highway collisions are a major cause of porcupine mortality—and most people only see porcupines in the form of roadkill. Perhaps that is why there are frequent comments about porcupines being "highway speed bumps."

DESCRIPTION: This large, stout-bodied rodent has long, white-tipped guard hairs surrounding the center of the back, where abundant, long, thick quills criss-cross one another in all directions.

RANGE: The Common Porcupine is widely distributed across most of Alaska and Canada, south to Pennsylvania and New England in the East and south through most of the West into Mexico.

DID YOU KNOW?

The name "porcupine," which comes from the Vulgar Latin porcospinus *(spiny pig), underwent many variations—Shakespeare used the word "porpentine"—before settling on its current spelling in the 17th century.*

Total Length: 21–37 in (53–94 cm)

Tail Length: 5½–9 in (14–23 cm)

Weight: 2¼–26 lb (1–12 kg)

The young are mostly black, but adults are variously tinged with yellow. The upper surface of the powerful, thick tail is amply supplied with dark-tipped, white to yellowish quills. The front claws are curved and sharp. The skin on the soles of the feet is covered with tooth-like projections. There may be gray patches on the cheeks and between the eyes.

HABITAT: Porcupines inhabit montane forests, foothills and even prairie environs in the mountains.

FOOD: Completely herbivorous, the Common Porcupine is like an arboreal counterpart of the American Beaver. It eats leaves, buds, twigs and especially young bark or the cambium layer of both broadleaf and coniferous trees and shrubs. During spring and summer, it eats considerable amounts of herbaceous vegetation. The porcupine typically puts on weight during spring and summer and loses it during fall and winter. It seems to have a profound fondness for salt, and it will chew and devour wood handles, boots and other material that is salty from sweat or urine.

DEN: Porcupines prefer to den in caves or shelters along watercourses or beneath fallen rocks, but they sometimes move into abandoned buildings, especially in winter. They are typically solitary animals, denning alone, but they may share a den during particularly cold weather. Sometimes a porcupine will sleep in a treetop for weeks, avoiding any den site, while it completely strips the tree of bark.

YOUNG: The porcupine's impressive armament inspires many questions about how it manages to mate. The female does most of the courtship, although males may fight with one another, and she is apparently stimulated by having the male urinate on her. When she is sufficiently aroused, she relaxes her quills and raises her tail over her back so that mating can proceed. Following mating in November or December and a gestation period of 6½ to 7 months—unusually long for a rodent—a single precocious porcupette is born in May or June. The young porcupine is born with quills, but they are not dangerous to the mother—the baby is born headfirst in a placental sac with its soft quills lying flat against its body. The quills harden within about an hour of birth. Porcupines have erupted incisor teeth at birth, and although they may continue to nurse for up to four months, they begin eating green vegetation before they are one month old. Porcupines become sexually mature when they are 1½ to 2½ years old.

walking trail

SIMILAR SPECIES: No other animal in the Rocky Mountains closely resembles the Common Porcupine, but there is a small chance that, in a nocturnal sighting, the Common Raccoon (p. 118) could be mistaken for a porcupine.

Common Raccoon

Meadow Jumping Mouse

Zapus hudsonius

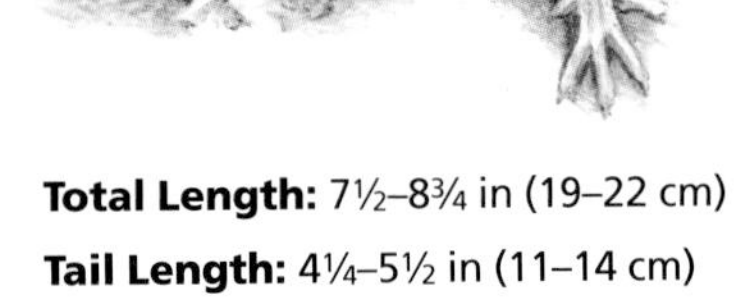

DESCRIPTION: The back is brownish, the sides are yellowish, and the belly is whitish. Juveniles are much browner dorsally than adults. The long, naked tail is dark above and pale below. The hindfeet are greatly elongated.

Total Length: 7½–8¾ in (19–22 cm)

Tail Length: 4¼–5½ in (11–14 cm)

Weight: ½–⅞ oz (14–25 g)

On the rare occasions when these fascinating mice are encountered, their method of escape belies their true identities: startled from their sedgy homes, jumping mice hop away in a manner befitting a frog. Unfortunately, this rodent's speed and the abundance of hideouts prevent extended observations.

Jumping mice spend up to nine months of the year in hibernation, which is not surprising, given that they rarely venture far from open water—they inhabit a landscape that is almost wholly frozen for much of the year. Adults are underground by the end of August; only those few juveniles that are below their minimum hibernation weight are active until mid-September.

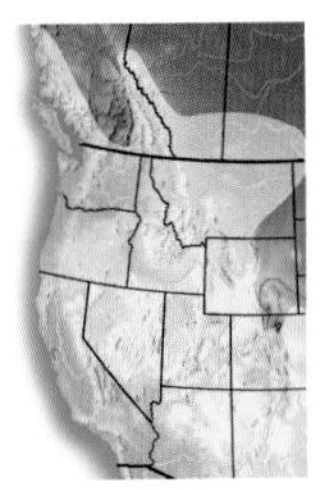

RANGE: This jumping mouse is found from southern Alaska across most of southern Canada and south to northeastern Oklahoma in the West and northern Georgia in the East.

HABITAT: Moist fields are preferred, but this jumping mouse also occurs in brush, marshes, brushy fields or even woods with thick vegetation.

FOOD: In spring, insects account for about half the diet. As the season progresses, the seeds of grasses and many forbs are eaten as they ripen. In summer and fall, subterranean fungi form about an eighth of the diet.

DEN: The Meadow Jumping Mouse hibernates in a nest of finely shredded vegetation in a burrow or other protected site. Its summer nest is built on the ground or in a small shrub.

YOUNG: These jumping mice breed within a week of the female's emergence from hibernation, typically in May. She bears two to nine young after a 19-day gestation. An extended, slow maturation follows: the eyes open after two to five days, and nursing continues for a month. The young must achieve a certain minimum weight or they will not survive the lengthy hibernation. Most females have a second litter after the first one leaves.

SIMILAR SPECIES: The Western Jumping Mouse (p. 157) has white edges on its ears and generally has a more southern range.

Western Jumping Mouse

Zapus princeps

Total Length: 7¾–10 in (20–25 cm)

Tail Length: 4¾–6 in (12–15 cm)

Weight: 11/16–1⅛ oz (19–32 g)

Let's face it, the Rockies' two species of jumping mice are virtually indistinguishable from each other in the field, which may create some confusion in areas where their ranges overlap. Their hopping escapes and supremely long tails, however, are sufficiently distinctive for even novice naturalists to distinguish either of them from most other mountain rodents.

DESCRIPTION: A broad, dark, longitudinal band extends from the nose to the rump. This dorsal stripe is primarily clay colored, with some blackish hairs. The sides of the body are yellowish olive, often with some orangish hairs. The belly is a clear creamy white. The flanks and cheeks are golden yellow. The naked tail is olive brown above and whitish below. The hindfeet are greatly elongated.

HABITAT: This jumping mouse prefers areas of tall grass, often near streams, that may have brush or trees. In the mountains, it ranges from valley floors up to treeline, and even into tundra sedge meadows. It frequently enters the water and appears to swim well, diving as deep as 3½ ft (1.1 m).

FOOD: In spring and summer, berries, tender vegetation, insects and a few other invertebrates are eaten. As fall approaches, grass seeds and the fruits of forbs are taken more frequently. Subterranean fungi are also favored.

DEN: The hibernation nest, made of finely shredded vegetation, is 1–2 ft (30–61 cm) underground in a burrow that is 3½–10 ft (1.1–3 m) long. The breeding nest is typically built among interwoven, broad-leaved grasses or in sphagnum moss in a depression.

YOUNG: Breeding takes place within a week after the female emerges from hibernation. Following an 18-day gestation period, four to eight young are born in late June or early July. The eyes open after two to five days. The young nurse for one month. Some females have two or even three litters a year.

SIMILAR SPECIES: The Meadow Jumping Mouse (p. 156) lacks the black hairs in the dorsal stripe and generally has a more northern range.

RANGE: This western species is found from the southern Yukon southeast to North Dakota and south to central California and northern Mexico.

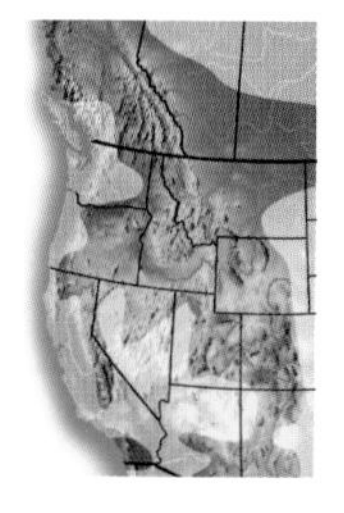

Western Harvest Mouse

Reithrodontomys megalotis

In contrast to the comparative ferocity of many mice, the Western Harvest Mouse has an angelic disposition. It is very tolerant of crowding, and individuals are known to huddle together to conserve heat in winter (the Western Harvest Mouse does not hibernate). This diminutive mouse appears to be inoffensive not only toward other harvest mice, but even toward other species. A female harvest mouse will tolerate having a male in her nest, and even strange mice may be introduced to the group without incident.

A leading candidate for the title of the smallest Rocky Mountain rodent, the Western Harvest Mouse is most active during the two hours after sunset, but its activity may continue almost until dawn, particularly on dark, moonless nights. It often uses vole runways through thick grass to reach foraging areas, and it is named for its habit of collecting grass cuttings in mounds along its trail networks. The Western Harvest Mouse does not store food in any great quantities, however, which is understandable for an animal that usually lives for less than a year.

Since the retreat of the Wisconsonian ice sheets, the Western Harvest Mouse has been expanding its range northward—mostly along either side of the Rocky Mountains—and eastward from the deserts of northern Mexico. It is among the rarest of our mammals: even in the more southern parts of its range, the Western Harvest Mouse is rarely as plentiful as other native mice.

DESCRIPTION: This native mouse closely resembles the House Mouse: it is small and slim, with a small head and pointed nose, and it has a conspicuous, long, sparsely haired tail and large, naked ears. The bicolored tail is grayish above and lighter below. The sleek upperparts are brownish, darkest down the middle of the back and on the ears. The underparts are grayish white, sometimes with a pale cinnamon wash, and they blend imperceptibly into the buffy tones on the cheeks, flanks and sides.

HABITAT: Harvest mice occur in both arid and moist places—grasslands, sagebrush, weedy waste areas, fencelines and even cattail-choked marsh edges—as long as there is abundant overhead cover.

FOOD: This harvest mouse eats lots of green vegetation in spring and early summer, and at those times of year its

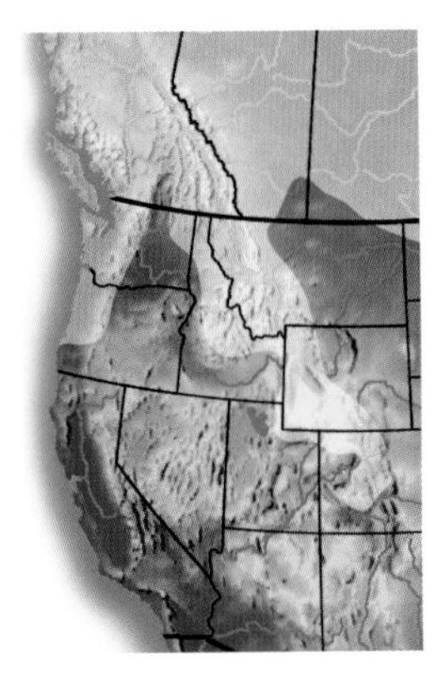

RANGE: This mouse ranges from extreme southern Alberta and British Columbia south nearly to the Yucatan peninsula of Mexico. It does not inhabit the roughest parts of the central mountains.

DID YOU KNOW?

The reproductive potential of harvest mice is remarkable, although high densities of the species have never been reported. Two recorded females produced 14 litters in 11 and 12 months respectively, with the total numbers of young being 57 and 58.

Total Length: 4¼–6 in (11–15 cm)

Tail Length: 2⅜–3⅛ in (6–7.9 cm)

Weight: 5⁄16–⅞ oz (8.9–25 g)

runways may be lined with piles of grass cuttings. During most of the year, however, the seeds of grasses, legumes, mustards and grains dominate the diet. Many grasshoppers, beetles, weevils and green sedges are also eaten throughout the year. The Western Harvest Mouse apparently does not store food.

DEN: This mouse builds its ball-shaped nest, which is about 3 in (7.6 cm) in diameter, either on the ground or low in a shrub or weeds. It constructs the nest with grass and other plant fibers and usually lines the central chamber, which is about 1½ in (3.8 cm) across, with finer downy material, such as cattail or milkweed fluff. The nest has one or more entrances, each about ⅜ in (1 cm) across, on its lower surface. A single nest may house 2 to 10 of these docile mice.

YOUNG: Reproduction is concentrated in the warmer months, but if conditions are right adult females may be almost continuously pregnant. The average litter of four is born after a 23- to 24-day gestation. At 4 days the incisors erupt; the hair is visible by 5 days; the eyes open after 10 to 12 days; and at 19 days the young are weaned. A female becomes sexually mature at four to five months.

SIMILAR SPECIES: The House Mouse (p. 174) is generally larger, its tail is gray-brown, it has a faint, black mid-dorsal stripe from head to tail, and it does not have the buffy tones on the cheeks and flanks.

Deer Mouse

Peromyscus maniculatus

To even the most committed mouse-o-phobe, the Deer Mouse looks cute. The large, protruding, coal black eyes give it a justifiably inquisitive look, while its dainty nose and long whiskers continually twitch, sensing odorous changes in the wind.

Wherever there is ground cover, from thick grass to deadfall, Deer Mice scurry about with great caution. These small mice are omnipresent over much of their range, and they may well be the most numerous mammal in the Rocky Mountains. When you walk through forested wilderness areas, they are in your company, even if their presence remains hidden.

Deer Mice most frequently forage along the ground, commuting between piles of ground debris, but they are known to climb trees and shrubs to reach food. During winter, Deer Mice are the most common of the small rodents to travel above the snow. In doing so, however, characteristically bounding along and leaving four neat little footprints, Deer Mice are vulnerable to nighttime predators. The tiny skulls of these rodents are among the most common remains in the regurgitated pellets of owls, which is a testament to their importance in the food web.

The Deer Mouse, which is named for the similarity of its coloring to that of the White-tailed Deer, commonly occupies farm buildings, garages and storage sheds, often alongside the House Mouse. There have been a few high-profile cases of people dying from the Hanta virus, which can be associated with the feces and urine of the Deer Mouse contaminating human food. The virus can become airborne, so if you find Deer Mouse droppings, it is best to wear a mask and spray the area with water and bleach before attempting to remove the animal's waste.

DESCRIPTION: All Deer Mice have large ears, a pointed nose, long whiskers, bright white undersides and feet, a cluster of whitish hairs at the front of each ear base, protruding, black, lustrous eyes and sharply bicolored tails with a dark top and light underside. In contrast to these constant characteristics, the color of the adult's upperparts is quite variable: yellowish buff, tawny brown, grayish brown or blackish brown. A juvenile has uniformly gray upperparts.

HABITAT: These ubiquitous mice occur in a variety of habitats, including prairie grasslands, mossy depressions, brushy

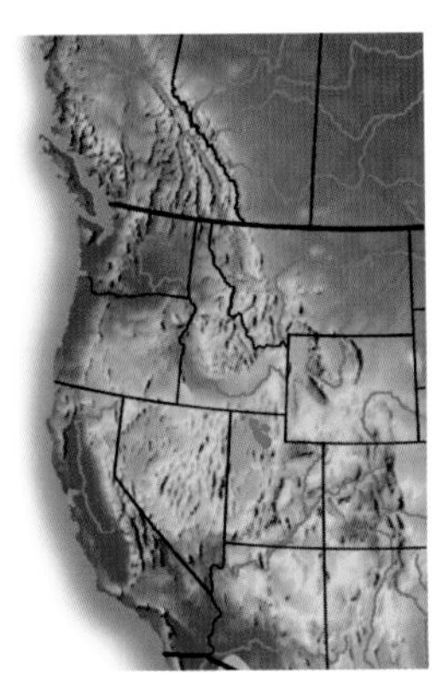

RANGE: The Deer Mouse is the most widespread mouse in North America. Its range extends from Labrador almost to Alaska and south through most of North America to south-central Mexico.

DID YOU KNOW?

Adult Deer Mice displaced a mile from where they were trapped were generally able to return to their home burrows within a day. Perhaps they range so widely in their travels that they recognized where they were and simply scampered home.

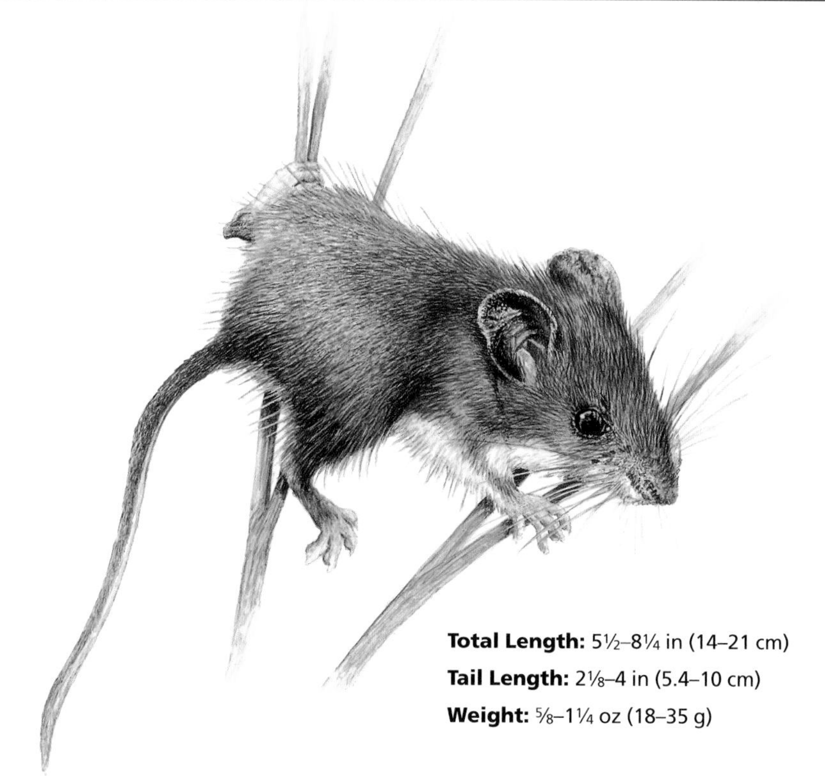

Total Length: 5½–8¼ in (14–21 cm)

Tail Length: 2⅛–4 in (5.4–10 cm)

Weight: ⅝–1¼ oz (18–35 g)

areas, tundra and heavily wooded regions. Another habitat that these little mice have a profound tendency to enter is the human building—our warm, food-laden homes are palatial residences to Deer Mice.

FOOD: Deer Mice use their internal cheek pouches to transport large quantities of seeds from grasses, grains, chokecherries, buckwheat and other weeds to their burrows. They also eat insects.

DEN: As the habitat of this mouse changes, so does its den type: in prairies and meadows it nests in a small burrow or makes a grassy nest on raised ground; in wooded areas it makes its nest in a hollow log or under forest-floor debris. The Deer Mouse may also nest in rock crevices, and certainly in manmade structures.

YOUNG: Breeding takes place between March and October, and gestation lasts for three to four weeks. The helpless young number one to nine (usually four or five) and weigh about 1/16 oz (1.8 g) at birth. They open their eyes between days 12 and 17, and about four days after that they venture out of the nest. At three to five weeks the young are completely weaned and are soon on their own. A female is sexually mature in about 35 days; a male in about 45 days.

SIMILAR SPECIES: Other *Peromyscus* mice (pp. 162–65) generally have longer tails, but habitat and range are often the best distinguishing features. Jumping mice (pp. 156–57) have much longer tails. The House Mouse (p. 174) and the Western Harvest Mouse (p. 158) lack the distinct white belly. The Bushy-tailed Woodrat (p. 170) is much larger.

Canyon Mouse

Peromyscus crinitus

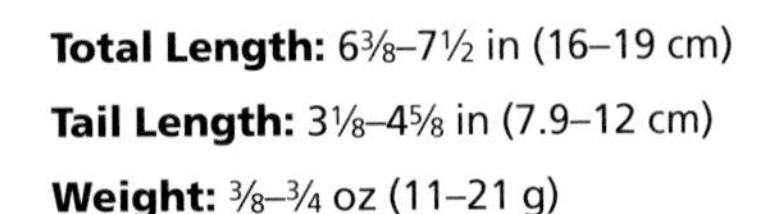

Total Length: 6⅜–7½ in (16–19 cm)

Tail Length: 3⅛–4⅝ in (7.9–12 cm)

Weight: ⅜–¾ oz (11–21 g)

"How dry I am" would be a most appropriate motto for the Canyon Mouse. This curious and endearing rodent has a particularly effective metabolism for life in arid regions. It rarely drinks—it can survive on the water it metabolizes from its food—and in situations of extreme aridity and heat it rests in torpor until conditions improve. Because of this tie to dry conditions, the distribution of old remains of this animal can hint to past climatic conditions. Canyon Mouse bones dating to the 13th century were found to the east and southeast of its present range, which may indicate that conditions were drier there at that time.

RANGE: The Canyon Mouse is found west of the continental divide, from eastern Oregon south through eastern California and east to western Colorado and northwestern New Mexico.

DESCRIPTION: The Canyon Mouse has very large, sparsely furred ears. Its thinly haired tail is longer than the head and body and distinctively bicolored. The underparts and feet are white. The upperparts are brown.

HABITAT: Bare rock seems to be a habitat requirement of the Canyon Mouse. It is found on desert pavement or bare canyon walls where strands of black brush, saltbrush, bunchgrass and sagebrush grow.

FOOD: These mice have quite a diverse, omnivorous diet that changes seasonally between seeds and other vegetation and insects.

DEN: Nests are located in burrows, among rocks, in logs or buildings or in other protected areas.

YOUNG: Typically, the Canyon Mouse, which lives in higher elevations, has two litters a year. Breeding occurs in early spring and early fall, and the average litter size is one to five young.

SIMILAR SPECIES: The Pinyon Mouse (p. 164) tends to have larger ears. The Deer Mouse (p. 160) tends to have a shorter tail. Habitat and range are often the best indicators of species identity.

Brush Mouse

Peromyscus boylii

Total Length: 7⅛–9⅜ in (18–24 cm)

Tail Length: 3⅝–4¾ in (9.2–12 cm)

Weight: ¾–1¼ oz (21–35 g)

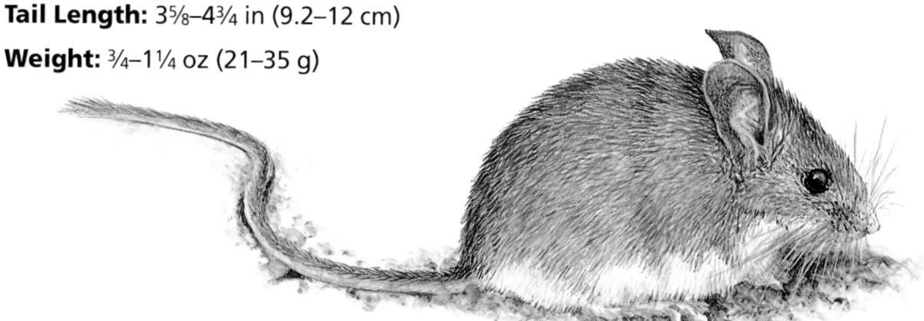

Brush Mice take a slightly different approach to evading danger. Rather than taking to the undergrowth like most frightened mice, they earn their name by commonly going vertical. Brush Mice are not master arborealists in the league of tree squirrels, but, nevertheless, their attempts at climbing often successfully separate them from their threats. The long tail thrashes back and forth, counterbalancing the animal like a tightrope walker with a balance pole.

These mice commonly occur in evergreen oak communities with lots of underbrush in the southern U.S. Rockies. Areas without underbrush are ill-suited to this mouse and are occupied by other types of mice, such as the Pinyon Mouse.

DESCRIPTION: This small to medium-sized mouse has a brown back and a sharply demarcated white belly. The feet are white, the ears are large and the long tail is bicolored. There is a rather indistinct tuft on the tail tip.

HABITAT: The Brush Mouse can be found in rocky and arid areas, particularly where there is some oak, juniper or pinyon vegetation.

FOOD: Brush Mice have been known to eat a wide variety of food items, many of which are only seasonally available. Seeds of conifers, berries and insects are common items, while the fruit of cactus is also consumed.

DEN: Nests are located in burrows, among rocks, in logs or buildings or in other protected areas.

YOUNG: Although it is capable of reproducing every month of the year under ideal situations, in the Rocky Mountains there are likely only two litters a year. Breeding likely takes place between March and May, and then again in October. Litters contain one to five (usually three) young.

SIMILAR SPECIES: The Deer Mouse (p. 160) has a tail that is shorter than the length of the head and body. The Pinyon Mouse (p. 164) has longer ears. The Canyon Mouse (p. 162) has shorter ears.

RANGE: The Brush Mouse is a western animal that ranges from Oregon, Utah and Colorado south through northern Mexico.

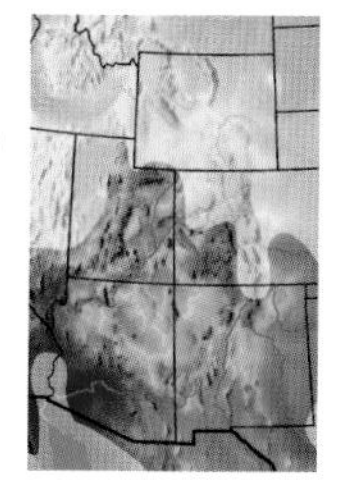

Pinyon Mouse

Peromyscus truei

Total Length: 6¾–9⅛ in (17–23 cm)

Tail Length: 3–4⅞ in (7.6–12 cm)

Weight: ⅝–1⅛ oz (18–32 g)

Along the southern toe of the Rocky Mountain chain, a delicate mixture of desert animals combines with those of the northern mountains. There, in the pinyon-juniper woodlands, there is one mammal that is more common than all the rest: the Pinyon Mouse. Tucked passively away during the frying heat of the day, it emerges from beneath rocks, shrubs and deadfall to scour the ground from dusk to dawn. Although it is not active during the hottest part of the day, the Pinyon Mouse can be detected by its sign—discarded pinyon and juniper seed husks set in loose piles tell the story of this mouse's food preferences.

DESCRIPTION: The back varies from lead colored to brownish, cinnamon or rich tawny, and it is separated sharply from the white to creamy white underparts. The thin tail is distinctly hairy and bicolored: dark above and white below. The ears are large and sparsely haired, the eyes are large and protruding, the nose is pointed and the whiskers are long. The feet are white.

HABITAT: The Pinyon Mouse is an inhabitant of arid foothill lowlands, and it is seldom found above 7000 ft (2130 m). It seems restricted to areas of pinyon and juniper, particularly where rocky slopes dominate.

FOOD: During summer, most of the diet consists of insects and spiders, while from late summer to spring, seeds and nuts, particularly from junipers and pinyons, are important.

DEN: Nests are located among rocks, in logs and hollow trunks, in buildings or in other protected areas.

YOUNG: The Pinyon Mouse has several litters between April and September, each consisting of three to six young.

SIMILAR SPECIES: The Canyon Mouse (p. 162) has smaller ears and inhabits rocky areas.

RANGE: The Pinyon Mouse ranges from central Oregon to Colorado and south through Mexico.

Northern Rock Mouse

Peromyscus nasutus

Total Length: 7⅛–10 in (18–25 cm)
Tail Length: 3⅝–5⅝ in (9.2–14 cm)
Weight: ⅞–1⅛ oz (25–32 g)

Well-suited to the rocky environment of the mountains, the Northern Rock Mouse uses its sharp claws and strong legs to scamper skillfully along cliff faces and rock walls. This mouse closely resembles a Deer Mouse, but it has a much longer tail, and it is larger and better at handling rocky terrain. It forages throughout the night over its home territory, which can be as large as several acres. The Northern Rock Mouse stores small amounts of food, but its caching behavior is not nearly as strong as that of other mice.

DESCRIPTION: The brownish back and the white belly are separated by a sharp line. The nose is pointed, the dark, lustrous eyes protrude noticeably, and the ears are large and rounded. The tail is more than half the length of the body, and it is brown on top and light below.

HABITAT: This mouse occurs in a wide variety of rocky terrain, such as talus slopes, cliffs, canyons and old lava formations, that is associated with pinyon and juniper trees.

FOOD: Much of this animal's diet is composed of pinyon nuts and juniper cones. In summer, it avidly consumes grasshoppers, beetles and other insects.

DEN: These mice nest in crevices in their rocky homes. A nest typically has accumulations of debris, soft vegetation or fur and food discards.

YOUNG: Very little is known about the reproductive cycle of this mouse, but presumably it is similar to that of the Deer Mouse. Breeding likely occurs in spring, and perhaps for a second time in fall. The litter size may be about one to five young, and the gestation about 25 days.

SIMILAR SPECIES: The Deer Mouse (p. 160) is smaller and usually yellower. The Pinyon Mouse (p. 164) is grayer and has a shorter tail. The Brush Mouse (p. 163) and Canyon Mouse (p. 162) have smaller ears.

RANGE: The Northern Rock Mouse is found from southeastern Utah and western Colorado south through eastern Arizona and western Texas into Mexico.

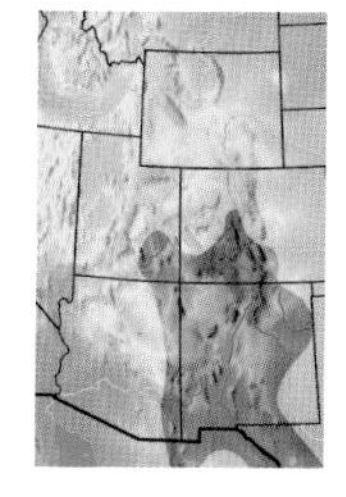

Northern Grasshopper Mouse

Onychomys leucogaster

Although it superficially resembles the Deer Mouse, the Northern Grasshopper Mouse is the bulldog of Rocky Mountain mice. It is a chunky resident of sandy habitats, and it often lives in close association with kangaroo rats. True to its stocky form, the Northern Grasshopper Mouse has a predatory nature—up to 90 percent of its diet consists of animals, primarily grasshoppers and other insects, but also including mammals as large as mice and voles.

This mouse is reputed to have a fierce disposition toward non-prey animals, as well. It frequently usurps the homes of other small mammals and modifies their burrows to suit its own needs. The grasshopper mouse locates its nest only a short distance down from the surface. Its burrow, whether it is second-hand or custom-built, is found in loose, dry, sandy soil, which is also ideal for dust bathing (a grasshopper mouse's form of sanitation to dress its naturally oily fur).

In contrast to its attitude toward strangers, the Northern Grasshopper Mouse seems to make a devoted parent. Both the male and female care for the young, bringing food to the nest until they become self-sufficient.

DESCRIPTION: The back is gray (in northern populations) to yellowish buff (in southern populations) and the belly is entirely white. The nose is pointed, the dark, lustrous eyes protrude noticeably, and the ears are large. The short tail—it is less than twice the length of the hindfoot—is thick, sharply bicolored (darker above, white below) and has a white tip. The thick legs, broad feet and broad shoulders of this mouse give it an overall impression of burliness. Animals seen in the wild may appear rumpled. The rumpled appearance and stink that grasshopper mice tend to develop is reduced by their sand bathing.

HABITAT: This mouse can be found in a wide variety of open habitats with sandy or gravelly soils, from grasslands to sandy brushlands, but it avoids alkali flats, marshy areas and rocky sites. Although it does not occur in the high mountains, it occupies the foothills in a few places.

FOOD: Only a little more than 10 percent of the summer diet consists of vegetation, mostly the seeds of grasses and forbs. Grasshoppers, crickets and beetles make up about 60 percent. In winter, up to 40 percent of the diet is

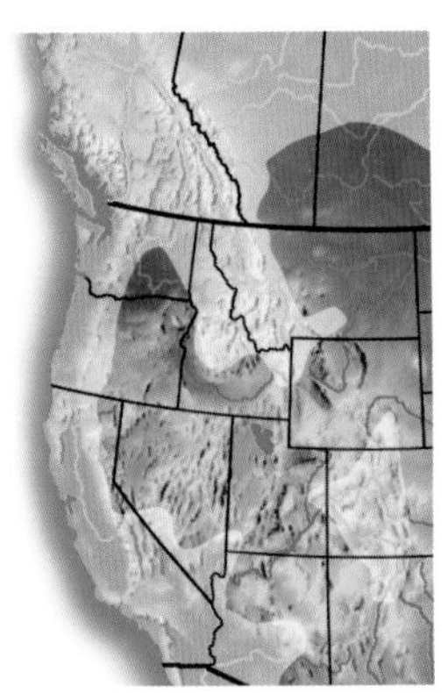

RANGE: This mouse ranges through much of the Great Plains, Rocky Mountains and Great Basin, from southern Alberta and Saskatchewan south into northern Mexico.

DID YOU KNOW?

The Northern Grasshopper Mouse has the ability to produce complex vocalizations. Mated pairs hunt simultaneously, and they apparently keep in contact with frequent, variable, bird-like calls that can be heard 100 yd (90 m) away.

Total Length: 5–6 in (13–15 cm)

Tail Length: 1⅛–1⅝ in (2.9–4.1 cm)

Weight: 11⁄16–1⅞ oz (19–53 g)

composed of seeds and vegetation. This fierce little predator may also eat scorpions, spiders, moths and butterflies. It typically bites off a scorpion's sting and then eats the animal tailfirst. A grasshopper mouse can even overpower and kill mice and birds that are up to three times its own weight. It does not appear to store food, and many grasshopper mice that are fat in fall lose weight over winter.

DEN: Nest burrows are U-shaped, and the tunnels are about 1½ in (3.8 cm) in diameter. The nest is located about 6 in (15 cm) below the surface. The burrow entrance is plugged by day to retain moisture. Nests may also be found under vegetation or debris, or in holes dug by other animals.

YOUNG: Grasshopper mice breed between March and August. A female's first pregnancy lasts about one month, but subsequent litters are typically born after a 32- to 38-day gestation. A litter usually contains three or four young, which weigh 1⁄16–⅛ oz (1.8–3.5 g) and are naked and blind at birth. The incisors begin to erupt at nine days, the eyes open at two to three weeks, and weaning follows by day 24. Most females breed during the spring following their birth, as do most males, and they may bear two or three litters each summer.

SIMILAR SPECIES: The Deer Mouse (p. 160) has a similar coloration but lacks the burly proportions and has a thinner, longer tail.

White-throated Woodrat

Neotoma albigula

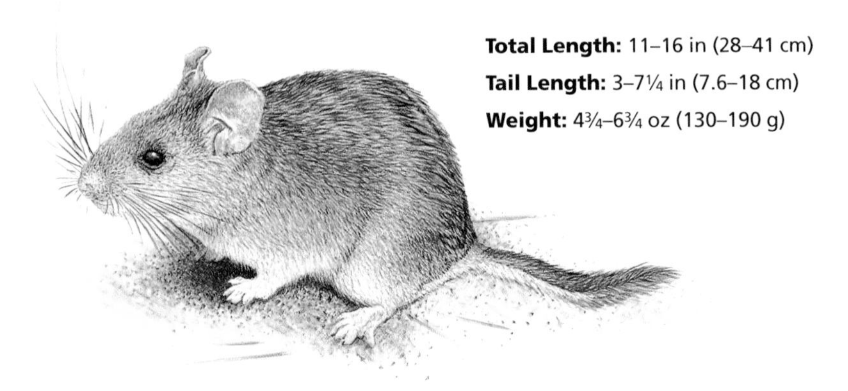

Total Length: 11–16 in (28–41 cm)

Tail Length: 3–7¼ in (7.6–18 cm)

Weight: 4¾–6¾ oz (130–190 g)

The White-throated Woodrat is an uncommon inhabitant of the Rocky Mountains. Only in the arid mountains of New Mexico and neighboring states is this woodrat found, and only in the deserts and brushy plains of that region does it reach a relatively high abundance. Appropriately, this woodrat is skilled at running uninjured over cacti. As with all woodrats, the White-throated Woodrat is an inquisitive and alert-looking animal that, when encountered, often shares a brief stare with a naturalist before it turns and retreats to safety.

DESCRIPTION: The upperparts are mainly brown or grayish, and the belly is white or light gray. The tail is lightly haired and colored the same as the body. The ears are moderately long, and the snout is pointed, with abundant, long whiskers. There is a patch of all-white hair at its throat.

RANGE: This woodrat is found in southern Colorado and Utah, most of Arizona and New Mexico into Texas and Mexico. Small numbers are found in extreme southeastern California.

HABITAT: A variety of arid habitats suit this woodrat, such as rocky terrain, scrublands, deserts and sagebrush flats. Usually the habitat includes cactus or pinyon and juniper.

FOOD: Cactus fruit appears to make up much of the diet. Others foods include juniper and pinyon fruits, yucca and some vegetation.

DEN: The nest is hidden with cactus spines and small twigs, and is usually at the base of a cactus or in a rocky crevice. Rarely does this woodrat use an underground chamber.

YOUNG: Breeding takes place from January to July. Females may have multiple litters of two or three young in a season. Their development is similar to that of other woodrats, with the young becoming sexually mature the spring after their birth.

SIMILAR SPECIES: The Mexican Woodrat (p. 169) lacks the all-white hair at the throat. The Bushy-tailed Woodrat (p. 170) has a distinctly bushier tail.

Mexican Woodrat

Neotoma mexicana

Total Length: 11–17 in (28–43 cm)

Tail Length: $4\frac{1}{8}$–$8\frac{1}{8}$ in (10–21 cm)

Weight: 5–$6\frac{1}{2}$ oz (140–180 g)

The Mexican Woodrat hardly does honor to the packrat name and tradition. This animal has weakened collecting instincts in comparison to the other woodrats of North America, and it doesn't belong at all in the same hoarding league as the Bushy-tailed Woodrat. Mexican Woodrats even seem capable of resisting the temptation of things that are shiny in favor of things that are practical. Much of the Mexican Woodrat's gathering is restricted to leafy green vegetation that is used either as food or in shelter construction.

DESCRIPTION: The back is predominantly grayish, but it is sometimes flecked with black or with orange highlights. The belly and feet are white or buffy white. A fringe of dark hairs surrounds the mouth. The lightly haired tail is black to gray on top and white below. The soles of the hindfeet are hairless to the heel. The ears are moderately long. The snout is pointed, and it has many long whiskers.

HABITAT: Rocky places within open stands of conifers are preferred. Sites with ponderosa pine, pinyon-juniper, scrub oak and mountain mahogany are suitable. This woodrat can be found to elevations of almost 8500 ft (2590 m).

FOOD: Green vegetation appears to make up most of the spring and early summer diet. In late summer and fall, leafy twigs are collected for winter use.

DEN: The nest of shredded, fibrous material is generally located well under rocks or sometimes in an old building. This woodrat does not regularly use large sticks, bones, cactus or debris in its nest.

YOUNG: Breeding takes place from late spring through summer. A female may have two litters of three or four young in a season. Their development parallels that of the Bushy-tailed Woodrat, with the young maturing sexually the spring after their birth.

SIMILAR SPECIES: The Bushy-tailed Woodrat (p. 170) has a bushier tail and hairy hindfeet. The White-throated Woodrat (p. 168) has white-based hairs at its throat.

RANGE: The Mexican Woodrat occurs from southern Colorado and southern Utah south into Mexico.

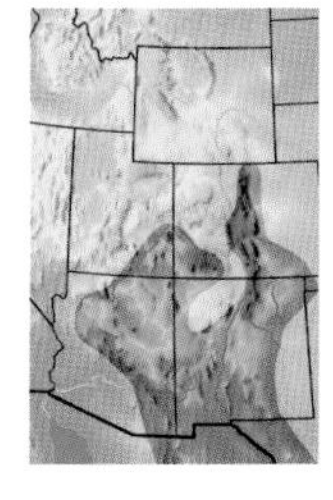

Bushy-tailed Woodrat

Neotoma cinerea

While most people have heard of "packrats," few people know that this nickname applies to woodrats. The Bushy-tailed Woodrat is a fine example of a packrat—it is widely known for its habit of collecting all manner of objects into a heap. This animal is also sometimes called a "trade rat," because it is nearly always carrying something in its teeth, only to drop that item to pick up something else instead. Thus, camping gear, false teeth, tools or even jewelry may disappear from a campsite, with a stick, bone or pinecone kindly left in its place.

Bushy-tailed Woodrats tend to nest in rocky areas, and because their nests are large and messy, woodrat homes are easier to find than the residents. The places in which woodrats can build their nests are limited, and rival males fight fiercely over their houses. Female woodrats are likely attracted to males that have secure nests, and several females may be found nesting with a single male.

The Bushy-tailed Woodrat may have proportionally the longest whiskers of any mammal in the Rockies. Extending well over the width of the animal's body on either side, a woodrat's whiskers serve it well as it feels its way around in the darkness of caves, mines and the night. Woodrats are most active after dark, so a late-night prowl with flashlights in hand may catch the reflective glare of woodrat eyes as the animals investigate their territories.

DESCRIPTION: The back is gray, pale pinkish or grizzled brown. The belly is white. The long, soft, dense, buffy fur is underlain by a short, soft underfur. The long, bushy, almost squirrel-like tail is gray above and white below. There are distinct juvenile and sub-adult pelages: a juvenile's back is gray, and it has short tail hairs; a subadult has brown hues in its back, and its tail has bushy guard hairs; a tawny adult pelage is developed in fall. All woodrats have large, protruding, black eyes, big, fur-covered ears and long, abundant whiskers.

HABITAT: This woodrat's domain usually includes rocks and shrubs or abandoned buildings, mine shafts or caves. It has a greater elevational range than other woodrats, extending from grasslands to alpine mountain regions.

FOOD: The leaves of shrubs are probably the most important component of the diet, but conifer needles and seeds,

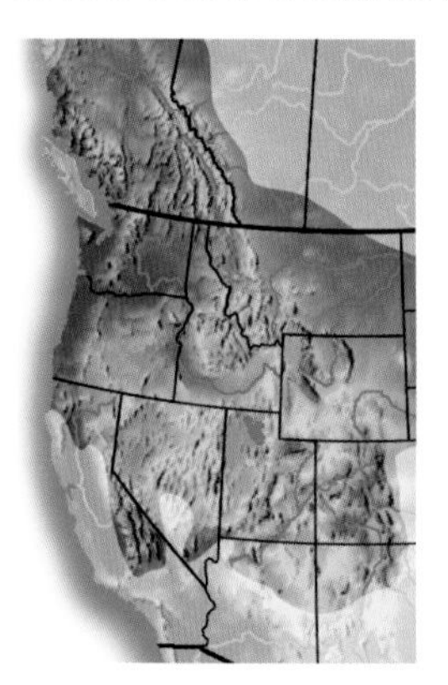

RANGE: The Bushy-tailed Woodrat is the most northerly woodrat. Its range extends from the southern Yukon southeast to western North Dakota and south to central California and northern New Mexico.

DID YOU KNOW?

When a very old woodrat nest in an old cabin near the Banff Springs Hotel was torn apart some years ago, a collection of hotel silverware dating back to the earliest days of the hotel was found.

Total Length: 11–18 in (28–46 cm)

Tail Length: 4⅜–8¾ in (11–22 cm)

Weight: 2¾–18 oz (78–510 g)

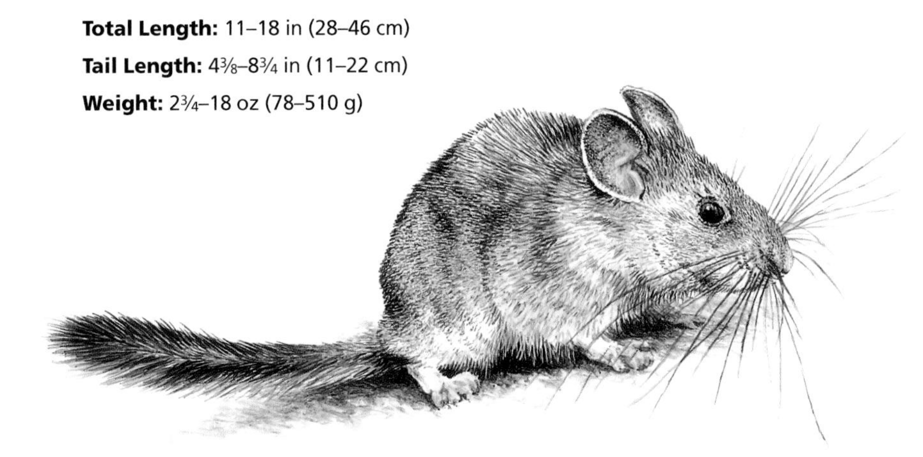

juniper berries, mushrooms, fruits, grasses, root stocks and bulbs are all eaten or stored for later consumption. To provide adequate winter supplies, a woodrat gathers and stores about 2 gal (7.6 *l*) of food. One woodrat may make several caches.

DEN: Large quantities of sticks, plus a large variety of bark, dung and other materials, are piled in a rock cleft or talus near the nest site. There are often no inner passages or chambers in this accumulation. Instead, a lined, ball- or cup-shaped nest is built of fibrous material and situated nearby, usually more than 11 ft (3.4 m) above the ground, either in a narrow crevice, in the fork of a tree, on a shelf or sometimes in a stove in an abandoned cabin.

YOUNG: Mating usually takes place between March and June. Following a 27- to 32-day gestation, three or four helpless young are born. They are 7⁄16–⅝ oz (12–18 g) at birth and their growth is rapid. Special teeth help them hold on to their mother's nipples almost continuously. Their incisors erupt at 12 to 15 days and the eyes open on day 14 or 15. They first leave the nest at about 22 days, and they are weaned at 26 to 30 days. The young reach sexual maturity the spring following their birth. Some females bear two litters in a season.

SIMILAR SPECIES: Other woodrats (pp. 168–69) look very similar, but none has as bushy a tail as this woodrat. The American Pika (p. 256) has no visible tail and its whiskers are dark.

walking trail

Norway Rat

Rattus norvegicus

While it is said that absence makes the heart grow fonder, it's a sure bet that no one misses rats. Most of the Rocky Mountain region is inhospitable to Norway Rats, but they may be found in areas of development and ranching. Everywhere Norway Rats occur, they are subject to public scorn and intense pest control measures.

The geography of the Rocky Mountains helps to limit the spread of rats throughout the region. Mountains are insurmountable to the rats, and the cold winters are fatal unless a warm building is nearby. The greatest influx of rats in the region arrives courtesy of modern transportation—animals hitch-hiking on trucks and trains are of concern, because they often get deposited in warm buildings in mountain cities and towns.

Norway Rats were introduced to North America in about 1775, and since then they have established colonies in most cities and towns south of the boreal forest. These great pests feed on a wide variety of stored grain, garbage and carrion, they gnaw holes in walls and they contaminate stored hay with urine and feces. Rats have also been implicated in the transfer of diseases to both livestock and humans.

More so than any other animal, Norway Rats are viewed with disgust by most people. They are described as filthy, loathsome creatures that eat vast quantities of stored food and live in the sewers, docks and warehouses of major cities worldwide. As one of the world's most studied and manipulated animals, however, much of our biomedical and psychological knowledge can be directly attributed to experiments involving this animal, which is a rather significant contribution for a hated pest.

This rat is capable of dispersing 3–5 mi (4.8–8 km) in a summer, but if it cannot find shelter in buildings or garbage dumps, winter temperatures of 0° F (–18° C) will prove fatal.

DESCRIPTION: The back is a grizzled brown, reddish brown or black. The paler belly is grayish to yellowish white. The long, round, tapered tail is darker above and lighter below and is sparsely haired and scaly. The prominent ears are covered with short, fine hairs. Occasionally, someone releases an albino, white or piebald Norway Rat that had been kept in captivity.

HABITAT: Norway Rats nearly always live in proximity to human habitation.

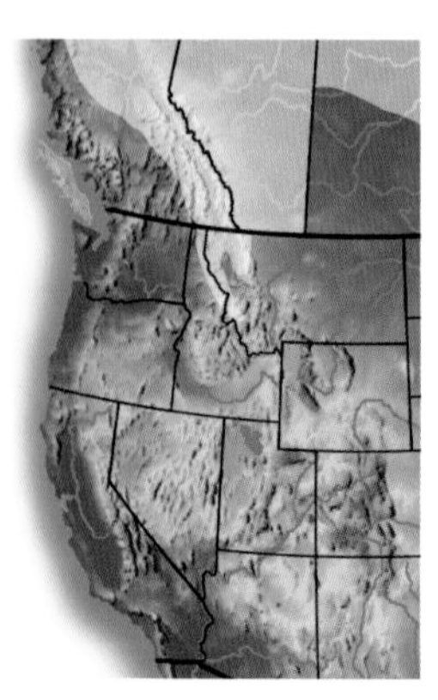

RANGE: The Norway Rat is concentrated in cities, towns and farms throughout coastal North America and south from southern Canada in the interior.

DID YOU KNOW?

Some historians attribute the end of the Black Death epidemics in Europe to the southward invasion of the Norway Rat and its displacement of the less aggressive Black Rat, which was much more apt to inhabit human homes.

Total Length: 13–18 in (33–46 cm)

Tail Length: 4¾–8¾ in (12–22 cm)

Weight: 7–17 oz (200–480 g)

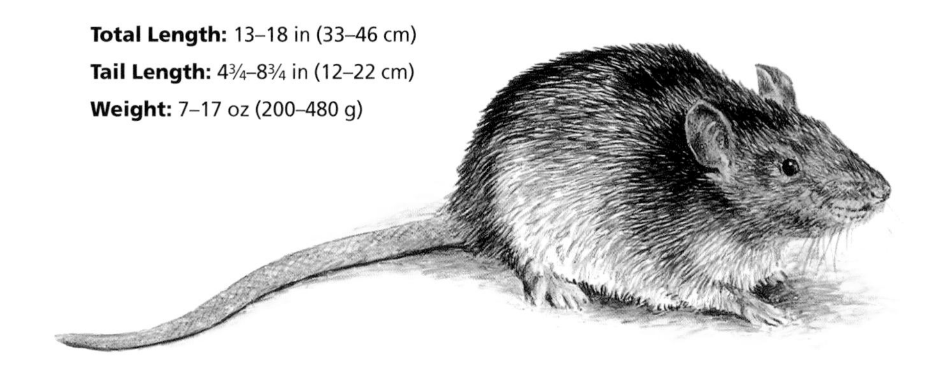

Where they are found away from humans, they prefer thickly vegetated regions with abundant cover. Abandoned buildings in the Rockies are more frequently occupied by Bushy-tailed Woodrats than by Norway Rats.

FOOD: This rat eats a wide variety of grains, insects, garbage and carrion; it may even kill young chickens, ducks, piglets and lambs. Green legume fruits are also popular items, and some shoots and grasses are consumed.

DEN: A cavity scratched beneath a fallen board or the space beneath an abandoned building may hold a bulky nest of grass, leaves and often paper or chewed rags. Although Norway Rats are able to, they seldom dig long burrows.

YOUNG: After a gestation of 21 to 22 days, 6 to 22 pink, blind babies are born. The eyes open after 10 days. The young are sexually mature in about three months. In the Rockies, Norway Rats seem to breed mainly in the warmer months of the year. Females may have several litters in a season.

SIMILAR SPECIES: The Bushy-tailed Woodrat (p. 170) has a white belly and its tail is covered with long, bushy hair. The Common Muskrat (p. 184) is generally larger and has a laterally compressed tail.

Black Rat

Rattus rattus

Total Length: 13–18 in (33–46 cm)

Tail Length: 6¼–10 in (16–25 cm)

Weight: 4¼–12 oz (120–340 g)

The Black Rat is generally restricted to the southern and coastal U.S., and the small numbers that are captured in the southern Rockies probably arrive via human transport. It is very similar in appearance to the Norway Rat, although generally slimmer and with a tail that is longer than the head and body.

House Mouse

Mus musculus

Thanks to its habit of catching rides with humans, first aboard ships and now in train cars, trucks and containers, the House Mouse is found in most countries of the world. In fact, the House Mouse's dispersal closely mirrors the agricultural development of humans. As humans began growing crops on the great sweeping plains of middle Asia, this mouse, native to that region, began profiting from our storage of surplus grains and our concurrent switch from a nomadic to a relatively sedentary lifestyle.

Within the small span of a few hundred human generations, farmed grains began to find their way into Europe and Africa for trade. Along with these grain shipments, stowaway House Mice were spread to every corner of the globe. Even in the Rocky Mountains, where most non-native mice and rats are unable to survive, the House Mouse is found wherever humans provide it free room and board.

House Mice are known to most people who have spent some time on farms or in warehouses, university labs and disorderly places. The white mice commonly used as laboratory animals are an albino strain of this species. Unlike many of the introduced animals in our region, House Mice seem to have had a minimal negative impact on our native animal populations.

DESCRIPTION: The back is yellowish brown, gray or nearly black, the sides may have a slight yellow wash, and the underparts are light gray. The nose is pointed and surrounded by abundant whiskers. There are large, almost hairless ears above the protruding black eyes. The long, tapered tail is hairless, gray and slightly lighter below than above. The brownish feet tend to be whitish on the terminal portion.

HABITAT: This widespread, introduced mouse inhabits homes, outbuildings, barns, granaries, hay stacks and trash piles. It cannot tolerate temperatures below 14° F (–10° C) around its nest, and it seems to be unable to survive winters in the northern forests without access to heated buildings or haystacks. In summer, some individuals may disperse slightly more than 2 mi (3.2 km) from their winter refuge into fields and prairies, only to succumb the following winter. Rocky Mountain cabins occupied by people only in summer are far more apt to be invaded by Deer Mice than by House Mice.

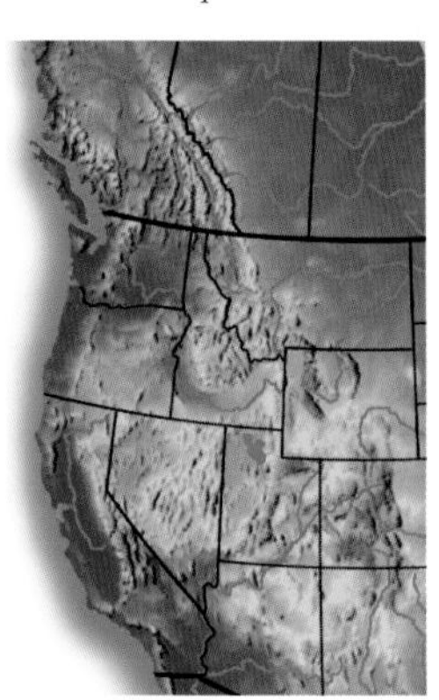

RANGE: The House Mouse is widespread in North America, inhabiting nearly every city, hamlet or farm from the Atlantic to the Pacific and north to the tundra.

DID YOU KNOW?

The word "mouse" probably derives from the Sanskrit mus*—also the source, via Latin, of the genus name—which itself came from* musha, *meaning "thief."*

Total Length: 5–7¾ in (13–20 cm)

Tail Length: 2½–4 in (6.3–10 cm)

Weight: ½–⅞ oz (14–25 g)

FOOD: Seeds, stems and leaves comprise the bulk of the diet, but insects, carrion and human food, including meat and milk, are eagerly consumed.

DEN: The nest is constructed of shredded paper and rags, vegetation and sometimes fur combined into a 4-in (10-cm) ball beneath a board, inside a wall, in a pile of rags or in a haystack. It may occur at any level in a building. House Mice sometimes dig short tunnels, but they generally do not use them as nest sites.

YOUNG: If abundant resources are available, as in a haystack, breeding may occur throughout the year, but populations away from human habitations seem to breed only during the warmer months. The gestation period is usually three weeks, but it may be extended to one month if the female is lactating when she conceives. The litter usually contains four to eight helpless, pink, jellybean-shaped young. Their fur begins to grow in two to three days, the eyes open at 12 to 15 days, and they are weaned at 16 to 17 days. At six to eight weeks, the young become sexually mature.

SIMILAR SPECIES: The Western Harvest Mouse (p. 158) looks very similar, but it has a clearly bicolored tail (lighter below) and a distinct longitudinal groove on the outside of each upper incisor tooth. The Deer Mouse (p. 160) has a bright white belly and distinctly bicolored tail.

Southern Red-backed Vole

Clethrionomys gapperi

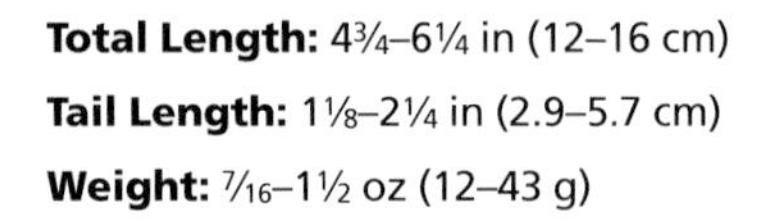

Total Length: 4¾–6¼ in (12–16 cm)

Tail Length: 1⅛–2¼ in (2.9–5.7 cm)

Weight: 7⁄16–1½ oz (12–43 g)

Although this handsome little vole, which is active both day and night, can be heard in the leaf litter of just about every sizable forest in the Rockies, it is almost never seen as it scurries along almost invisible runways on the forest floor.

The Southern Red-backed Vole is a classic example of a subnivean wanderer—a small mammal that lives out cold winters between the snowpack and the frozen ground. At ground level, snow changes into easily penetrated depth-hoar, forming a narrow, relatively warm layer within which this small rodent spends the winter. This vole does not even cache food; instead, it forages widely under the snow for vegetation or any other digestible foods.

DESCRIPTION: Reddish dorsal stripes make this animal one of the easiest voles to recognize. On rare occasions, the dorsal stripe is a rich brownish black or even slate brown. The sides are grayish buff, and the undersides and feet are grayish white. Compared to those of most voles, the black eyes seem small and the nose looks slightly more pointed. The short tail is slender and scantily haired. The rounded ears project somewhat above the thick fur.

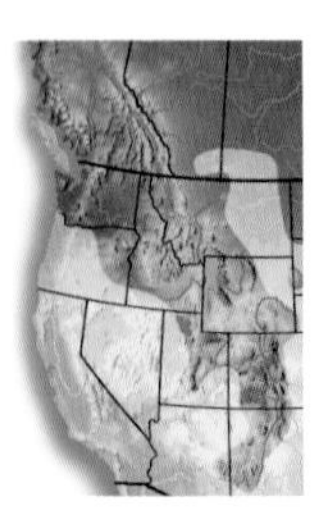

RANGE: This vole is widespread across southern Canada. It ranges south through the western mountains as far as New Mexico and through the Appalachians to North Carolina.

HABITAT: This vole is found in a variety of habitats, including damp and coniferous forests, bogs, the vicinity of swamps and sometimes drier aspen forests.

FOOD: Green vegetation, grasses, berries, lichens, seeds and fungi form the bulk of the diet.

DEN: Summer nests, made in shallow burrows, rotten logs or rock crevices, are lined with fine materials, such as dry grass, moss and lichens. Winter nests are subnivean—located above the ground but below the snow.

YOUNG: Mating occurs between April and October. Following a gestation period of about 20 days, two to eight (usually four to seven) pink, helpless young are born. They nurse almost continuously, and their growth is rapid. By two weeks they are well-furred and have opened their eyes. Once the young are weaned, they are no longer permitted in the vicinity of the nest. This vole reaches sexual maturity at two to three months.

SIMILAR SPECIES: The Long-tailed Vole (p. 182) has a longer tail, and both it and the Meadow Vole (p. 180) lack the reddish dorsal stripe.

Western Heather Vole

Phenacomys intermedius

Total Length: 4¼–6¼ in (11–16 cm)

Tail Length: 1–1⅝ in (2.5–4.1 cm)

Weight: ⅞–1¾ oz (25–50 g)

The Western Heather Vole usually occupies the alpine tundra, but it may descend to the same northern woodlands as its red-backed kin, and skulls of each species are not infrequently found in the same owl pellets.

This vole's common name is slightly misleading, because heather makes up little of either its dietary requirements or habitat preferences. "Bark vole" might have been a better label, because it consumes a high percentage of bark seasonally, and it has a cecum (a functional appendix) with modified, 3/8-in (1-cm) villi that assist in digesting this fibrous and lignin-rich food.

DESCRIPTION: This gentle vole has a short, thin, bicolored tail that is slate gray above, sometimes with a few white hairs, and white below. The tops of the feet are silvery gray, and the belly hairs have light tips, giving the entire undersurface a light gray hue. Various dorsal colors are seen, but the most common is a grizzled buffy brown. The ears are roundish and scarcely extend above the fur. There is tawny or orangish hair inside the front of the ear.

HABITAT: This vole seems to prefer open areas in a variety of habitats in the mountains, including alpine tundra and coniferous forests.

FOOD: This vole primarily feeds on green vegetation, grasses, berries, seeds, lichens and fungi. It has a strong tendency to eat the inner bark of various shrubs from the heather family.

DEN: The summer nest is made in a burrow up to 8 in (20 cm) deep, and it is lined with fine dry grass and lichens. In winter, the nest is built on the ground in a snow-covered runway.

YOUNG: Mating occurs between April and October. Following a gestation of about three weeks, one to eight (usually four or five) pink, helpless young are born. They nurse almost continuously and their growth is rapid. By two weeks, they are well furred and have opened their eyes. This vole becomes sexually mature after two to three months, but a young male usually does not breed in his first year.

SIMILAR SPECIES: The Long-tailed Vole (p. 182) has a longer tail, and both it and the Meadow Vole (p. 180) have slate gray hindfeet.

RANGE: This vole occurs from northwestern British Columbia south through the western mountains to central California and northern New Mexico.

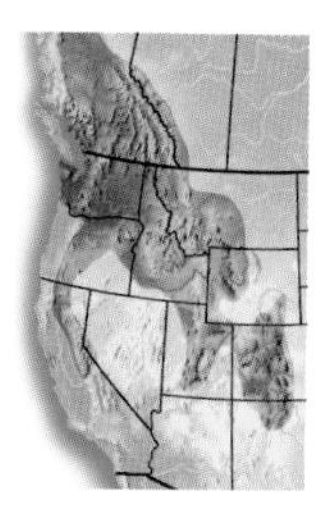

Water Vole

Microtus richardsoni

On hikes along high-elevation trails in the mountain parks, there is often the temptation to quickly cross the creeks that stream across your path. If you linger, however, you may have an opportunity to become familiar with the Water Vole.

This large vole is like a small alpine muskrat in many ways as it dives and forages with ease along the icy snowmelt creeks. It is almost exclusively nocturnal, unfortunately, so you are more likely to experience its distinctive sign than the animal itself.

The Water Vole's diagnostic, well-worn runways criss-cross the margins of alpine streams, connecting the burrows and foraging areas of small colonies. These damp pathways are often under the mat of roots and plant debris on the ground surface. Vegetation cuttings often line the paths. This semi-aquatic vole's burrows may be in such close proximity to the water that you would expect them to flood with each rainfall.

Water Voles appear to abandon their tunnel networks through the winter months, remaining adequately protected from the winter's chill by the snows that deeply coat and insulate their habitat. Few of them will survive more than one winter.

DESCRIPTION: The large Water Vole is brownish black above, with paler gray sides. The belly is gray with a grayish-white wash. The fur is thick, and there is an abundant, water-repelling undercoat. The extremely long hindfeet aid in swimming. The tail is indistinctly bicolored: the fur is blackish above and dark gray below. The ears are rounded, and they scarcely extend above the thick fur. The eyes are small, black and protruding.

HABITAT: True to its name, this vole lives primarily along alpine and subalpine streams and lakes. It favors clear, swift streams with gravelly bottoms that are lined with mixed stands of low willows and dense herbage.

FOOD: In summer, Water Voles feed on the shoots and culms of various sedges and grasses, as well as the roots, stems, leaves and flowers of forbs. Winter foods include the bark of willows and bog birch, various roots and rhizomes and the fruits and seeds of available green vegetation.

DEN: This vole digs extensive burrow systems, with tunnels up to 4 in (10 cm) in diameter, through moist soil at the

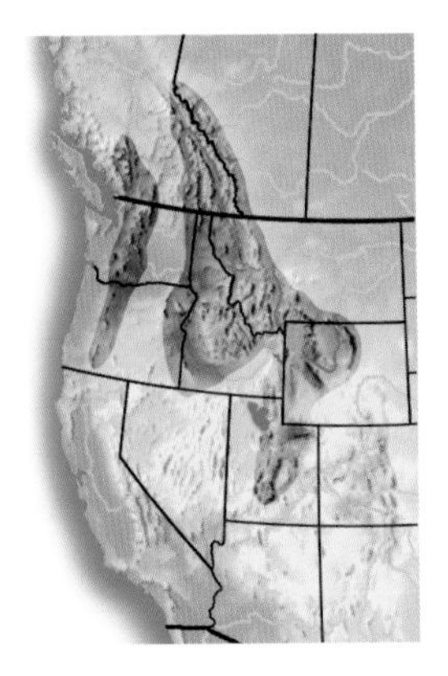

RANGE: The Water Vole has two disjunct populations, each associated with mountains: the western population extends along the Cascade Mountains from central British Columbia to southern Oregon; the eastern population occurs in the Rocky Mountains from western Alberta to south-central Utah.

DID YOU KNOW?

The Water Vole is an excellent swimmer, and it will often seek refuge from martens, weasels and other predators in the water.

Total Length: 7½–11 in (19–28 cm)

Tail Length: 2⅛–3⅞ in (5.4–9.8 cm)

Weight: 1⅛–4¼ oz (32–120 g)

walking trail

edges of streams or waterbodies. The nest chamber, which is about 4 in (10 cm) high and 6 in (15 cm) long, is lined with moss and dry grass or leaves. It is often situated under a rise, log or stump. A Water Vole will excavate and re-excavate its burrow system throughout the summer. The winter nest is located farther from the water in a snow-covered runway.

YOUNG: Water Voles probably breed periodically from May through September, with usually two or more litters of 2 to 10 young born each year. The gestation period is at least 22 days. The young are helpless at birth, but they grow rapidly, reaching maturity quickly. They may even breed during their birth year.

SIMILAR SPECIES: The Water Vole's large hindfoot, which is more than 1 in (2.5 cm) long, and generally large size distinguish it from all other voles in the Rockies.

Meadow Vole

Microtis pennsylvanicus

Total Length: 5–7½ in (13–19 cm)

Tail Length: 1¼–1¾ in (3.2–4.5 cm)

Weight: ⅝–2¼ oz (18–64 g)

When the snows recede from the land every April, an elaborate network of Meadow Vole activity is exposed to the world. Highways, chambers and nests, previously insulated from the winter's cold by deep snows, await the growth of spring vegetation to conceal them once again. These tunnels often lead to logs, boards or shrubs, where the voles can find additional shelter.

Many Meadow Voles die in their first months, and very few voles seem to live longer than a year. With two main reproductive cycles a year, it is unlikely that many voles get to experience all seasons. Despite their high mortality, Meadow Voles are often the most common animals in fields, owing to their explosive reproductive potential.

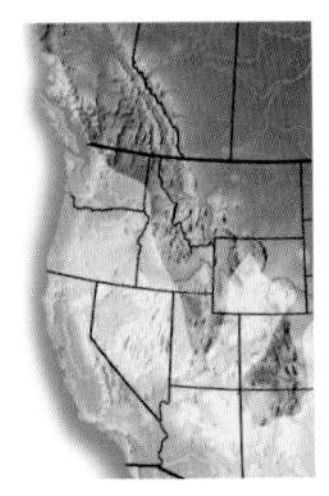

RANGE: Essentially a northern vole, it occurs from central Alaska to Labrador and south to Utah and northern New Mexico in the West and to Georgia in the East.

DESCRIPTION: The body is brown to blackish above and gray below. The protruding eyes are small and black. The rounded ears are mostly hidden in the long fur of the rounded head. The tops of the feet are blackish brown. The tail is about twice as long as the hindfoot.

HABITAT: With the broadest range of any vole in North America, the Meadow Vole can be found in a variety of habitats, including alpine tundra, taiga, deciduous forests, open plains, cultivated fields and around marshes, waterbodies and areas with dense herbaceous shrubs.

FOOD: The green parts of sedges, grasses and some forbs make up the bulk of the spring and summer diet. In winter, large amounts of seeds, some bark and insects are eaten. Other foods include ground beans, grains, roots and bulbs.

DEN: The summer nest is made in a shallow burrow and lined with fine materials, such as dry grass, moss and lichens. The winter nest is subnivean: above the ground but below the snow.

YOUNG: Spring mating typically occurs with the appearance of green vegetation between late March and the end of April. Gestation is about 20 days, and the average litter size is four to eight. From birth, the helpless young nurse almost constantly to support their rapid growth. Their eyes open in 9 to 12 days, and they are weaned at 12 to 13 days. At least one more litter is born, usually in fall.

SIMILAR SPECIES: The Montane Vole (p. 181) tends to be paler, and it is more common in the U.S. Rockies.

Montane Vole

Microtus montanus

Total Length: 5¼–7¼ in (13–18 cm)
Tail Length: 1¼–2¼ in (3.2–5.7 cm)
Weight: ½–1¾ oz (14–50 g)

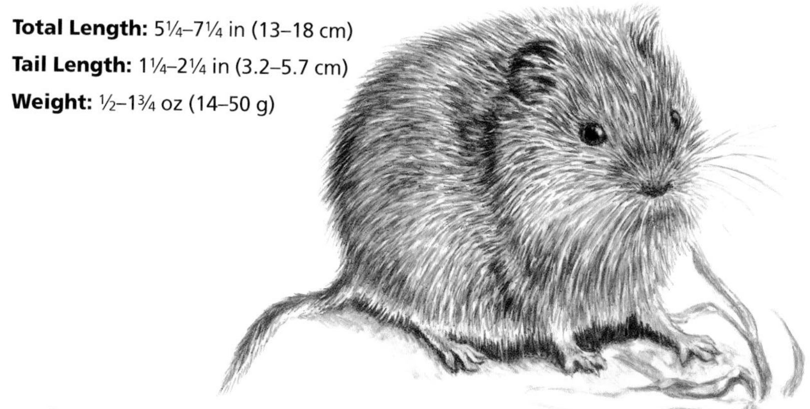

When the sun settles for the night and the moon casts its brightness over the Rocky Mountain forests, the sounds of the night do much to keep campers awake. Although the rustling of leaves and scurrying footsteps of mice and voles are the most common sounds, sleep comes slowly to campers with thoughts of other possibilities. Somehow, these faint noises intensify in our minds and convince us that bears, wolves and cougars are just outside the tent's fly. If Montane Voles were capable of feeling vindicated, this reaction would be sweet justice for these animals, which are overlooked by most visitors.

DESCRIPTION: The Montane Vole is a small, thickset mammal with short, largely hidden ears and dark protruding eyes. The back is brown to black. The belly is a lighter gray. The head is rounded and the snout is blunt. Most of each limb is hidden in the trunk's skin, giving it a short-legged appearance. The tail is comparatively long, bicolored and sparsely covered with hair.

HABITAT: This vole is found in alpine tundra and large, grassy, mountain meadows.

FOOD: Green, grassy shoots form the majority of the diet when they are available. At other times of the year, seeds or even bark may be eaten.

DEN: The nests are often located aboveground along well-used runways. The runways and nests are easiest to observe soon after the snow melts in spring.

YOUNG: Montane Voles may have several litters in a year. Reproduction usually takes place between spring and fall. The gestation is about 21 days, after which six to eight young are born.

SIMILAR SPECIES: The Meadow Vole (p. 180) looks extremely similar, but it tends to be darker, and it is less common in many parts of the U.S. Rockies.

RANGE: The Montane Vole occurs from British Columbia to Montana and south to New Mexico and California.

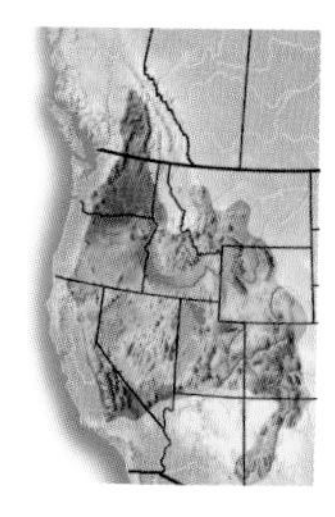

Long-tailed Vole

Microtis longicaudus

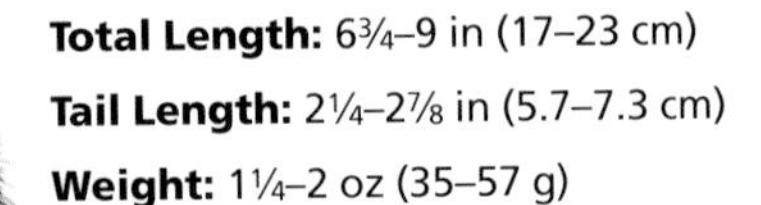

Total Length: 6¾–9 in (17–23 cm)

Tail Length: 2¼–2⅞ in (5.7–7.3 cm)

Weight: 1¼–2 oz (35–57 g)

Long-tailed Voles have a very odd distribution in the Rocky Mountains. These voles are among the alpine elite, thriving above treeline in the mountain parks, but they also live among their flatland kin on grassland plateaus. In both communities, Long-tailed Voles choose to live in wet meadows with stunted thickets. They do not follow well-defined trails, and they range widely at night.

DESCRIPTION: The upperparts are variously colored, ranging from grizzled grayish to dark gray-brown, but the black tips on the guard hairs may give this vole a dark appearance. The sides are paler than the back, and the undersides are paler still. The tail of this vole is indistinctly bicolored, and it is about 2⅜ in (6 cm) long. The uppersides of the feet are gray.

RANGE: The Long-tailed Vole ranges south from eastern Alaska and the Yukon along the Rocky Mountains to New Mexico and along the Pacific coast to California.

HABITAT: This vole lives in a variety of habitats, including dry grassy areas, mountain slopes, coniferous forests, alpine tundra and among alders or willows in the vicinity of water.

FOOD: Summer foods consist of green leaves, grass shoots, fruits and berries. In winter, this vole consumes the bark of heaths, willows and trees.

DEN: The simple burrows made under logs or rocks are often poorly developed. The nest chamber is lined with fine, dry grass, moss or leaves. The winter nests are subnivean.

YOUNG: Mating is presumed to occur from May to October, with the females often having two litters of two to eight (usually four to six) young a year. The gestation period is about three weeks. The young are helpless at birth, but at about the same time their eyes open at two weeks old, they are weaned and leave the nest. Some young females have their first litter when they are only six weeks old.

SIMILAR SPECIES: The Meadow Vole (p. 180) has a shorter tail and dark feet.

Sagebrush Vole

Lemmiscus curtatus

Total Length: 4¼–5½ in (11–14 cm)

Tail Length: 11⁄16–1⅛ in (1.8–2.9 cm)

Weight: ¾–1⅜ oz (21–39 g)

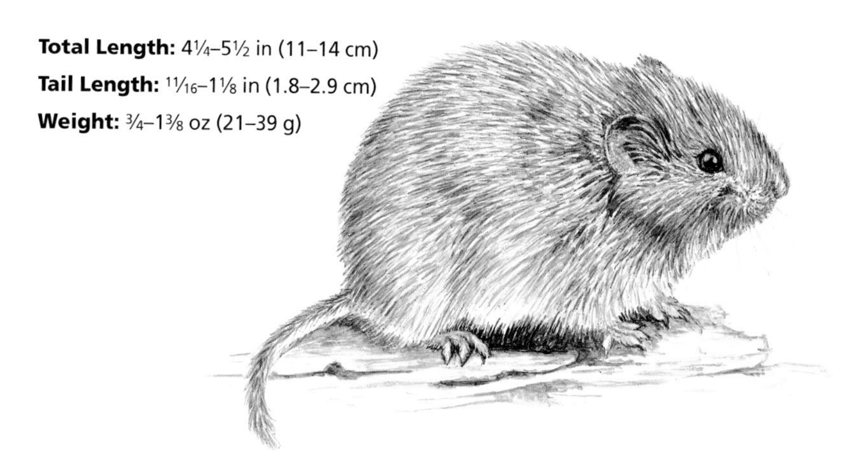

Decades of walking through prime Sagebrush Vole habitat, will, on very rare occasions, produce encounters with this secretive animal. Arid grass- and scrublands seem to offer little concealment for even this small rodent, but its pale colors help it blend in with the surroundings. Years of very low vole densities may also account for the difficulty in seeing one, but infrequent population irruptions can result in a 10-fold increase in numbers.

DESCRIPTION: This small, stout vole has short ears and legs and long, lax hair. It is a very pale ashy gray in color, with buffy tinges around the ears and nose. The undersides are silvery. The tail is not much longer than the hindfoot, and it is dark above and light below.

HABITAT: In keeping with its name, the Sagebrush Vole thrives in arid grassland regions and sagebrush flats.

FOOD: The spring and summer diets include a variety of plant and some insect material, especially the new green shoots of grasses and the flowers and leaves of forbs. In fall and winter, the diet switches to ripened fruits, seeds, bulbs, roots, corms and the inner bark of shrubs.

DEN: A shallow but extensive system of burrows leads to a grass-lined nest chamber.

YOUNG: Mating occurs mainly from April to September. After a gestation period of about 25 days, a litter of 1 to 13 helpless young is born. The young grow rapidly, and they are weaned and out of the nest within three weeks.

SIMILAR SPECIES: The Long-tailed (p. 182), Meadow (p. 180) and Montane (p. 181) voles all have tails that are longer than 2.5 cm (1 in).

RANGE: The range of this vole, which skirts much of western Montana and northern Idaho, extends from southern Alberta east to North Dakota, south to southern Nevada and north to central Washington.

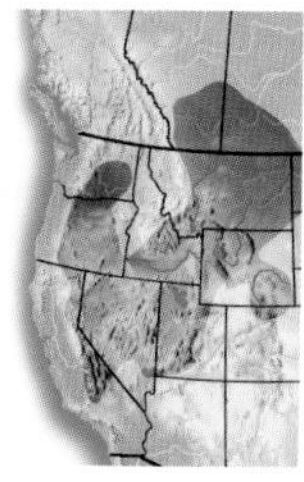

Common Muskrat

Ondatra zibethicus

After a long winter that restricts Common Muskrats to a life beneath the ice, the first few weeks of spring find many of these animals stretching their legs on land. It is usually in early May that many first-year animals, now sexually mature, venture from their birth ponds to establish their own territories. These muskrats are commonly seen traveling over land, which is a tragic requirement for many—their numbers can be all too easily tallied on May roadkill surveys.

The Common Muskrat is not a "mini-beaver," nor is it a close relative of that large rodent; rather, it is a highly specialized, aquatic vole that shares many features with the American Beaver as a result of their similar environments. Like a beaver, a muskrat can close its lips behind its large orange incisors so it can chew underwater without getting water or mud in its mouth. Its eyes are placed high on its head, and a muskrat can often be seen swimming with only its head and sometimes its tail abovewater. The Common Muskrat dives with ease; according to reports, it can submerge for 15 minutes and can swim the length of a football field before surfacing.

Muskrats lead busy lives. They are continuously gnawing cattails and bulrushes, whether eating the tender shoots or gathering the coarse vegetation for home building. Muskrat homes rise above shallow waters throughout the Rockies, and they are of tremendous importance not only to these aquatic rodents, but also to geese and ducks, which make use of muskrat homes as nesting platforms.

Both sexes have perineal scent glands that enlarge and produce a distinctly musk-like discharge during the breeding season. Although this scent is by no means unique to the Common Muskrat, its potency is sufficiently notable to have influenced this animal's common name. An earlier name for this species was "musquash," from the Abnaki name *moskwas*, but through the association with musk the name changed to "muskrat."

DESCRIPTION: The coat generally consists of long, shiny, tawny to nearly black guard hairs overlying a brownish-gray undercoat. The flanks and sides are lighter than the back. The underparts are gray, with some tawny guard hairs. The long tail is black, nearly hairless, scaly and laterally compressed with a dorsal and ventral keel. The legs are short. The hindfeet are large and

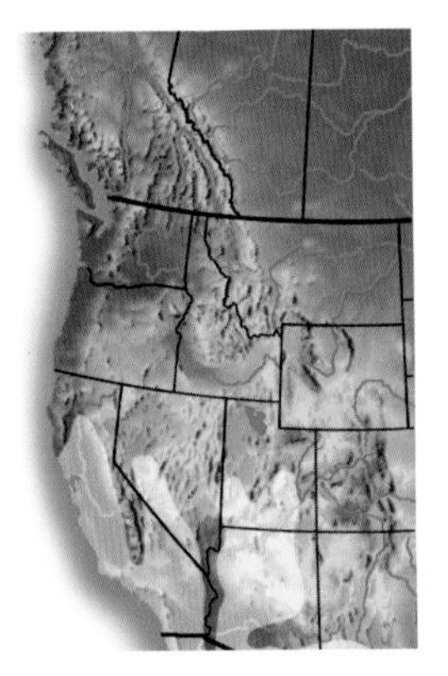

RANGE: This wide-ranging rodent occurs from the southern limit of the arctic tundra across nearly all of Canada and the lower 48 states except for most of Florida, Texas and California.

DID YOU KNOW?

Muskrats are highly regarded by native peoples. In one story, it was Muskrat who brought some mud from the bottom of the flooded world to the water's surface. This mud was spread over a turtle's back, thus creating all the dry land that we now know.

Total Length: 1½–2 ft (46–61 cm)

Tail Length: 7¾–11 in (20–28 cm)

Weight: 1¾–3½ lb (0.8–1.6 kg)

partially webbed and have an outer fringe of stiff hairs. The tops of the feet are covered with short, dark gray hairs. The claws are long and strong.

HABITAT: Muskrats occupy sloughs, lakes, marshes and streams that have cattails, rushes and open water. They are not present in the high mountains.

FOOD: The summer diet includes a variety of emergent herbaceous plants. Cattails, rushes, sedges, irises, water lilies and pondweeds are staples, but a few frogs, turtles, mussels, snails, crayfish and an occasional fish may be eaten. In winter, muskrats feed on submerged vegetation.

DEN: Muskrat houses are built entirely of herbaceous vegetation, without the branches or mud of beaver lodges. The dome-shaped piles of cattails and rushes have an underwater entrance. In places, muskrats may dig bank burrows, which are 15–50 ft (4.6–15 m) long and usually have underwater entrances.

YOUNG: Breeding takes place between March and September. Each female produces two or sometimes three litters a year. Gestation lasts 25 to 30 days, after which six to seven young are born. The eyes open at 14 to 16 days, the young are weaned at three to four weeks, and they are independent at one month old. Both males and females are sexually mature the spring after their birth.

SIMILAR SPECIES: The American Beaver (p. 188) is larger and has a broad, flat tail, and typically only its head is visible abovewater when it swims.

Brown Lemming

Lemmus trimucronatus

Total Length: 4–6¾ in (10–17 cm)

Tail Length: 7/16–1⅛ in (1.1–2.9 cm)

Weight: 1¾–4 oz (50–110 g)

The Brown Lemming is a colorful arctic furball that tolerates some of the most inhospitable environments in North America. In the Rockies, it is only known to occur in the alpine tundra of the most northern ranges. Every part of the body is covered with a long coat—most appropriate for an animal that typically lives in the tundra—that also provides the lemming with excellent buoyancy when it swims.

DESCRIPTION: The body, ears, feet, head and stubby tail are all completely covered with long, lax fur. In summer, the lower back is chestnut colored, grading to grizzled gray over the head and shoulders. The rump is a lighter brown, and the cheeks and sides are tawny. The undersides are primarily light gray. In fall, the lemming molts into a longer, grayer coat. The strong, curved claws aid in digging the elaborate winter runways.

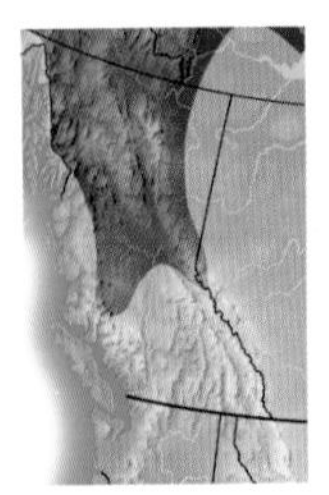

RANGE: The Brown Lemming primarily inhabits Alaska, the Yukon and the Northwest Territories. It ranges south into the mountains of northern British Columbia and Alberta.

HABITAT: Bogs, alpine meadows and tundra and even spruce woods may support large lemming colonies.

FOOD: Grasses, sedges and other monocotyledonous plants form the bulk of the diet. In times of scarce vegetation, any emergent plant is eaten down to the surface.

DEN: Summer nests of dry grass and fur are located 2–12 in (5.1–30 cm) underground in tunnels, with nearby chambers for wastes. Winter nests are subnivean: located above the ground but below the snow.

YOUNG: Breeding occurs from spring through fall, and sometimes in winter. Gestation is about three weeks, after which four to nine young are born. Lemmings resemble pink jelly beans at birth. Their growth is rapid: at 7 days they are furred, the ears open at 8 to 9 days, and the eyes open at 10 to 12 days. The young are weaned at 16 to 21 days, and they are probably sexually mature soon thereafter.

SIMILAR SPECIES: The Northern Bog Lemming (p. 187) is generally smaller and has a bicolored tail and grooved upper incisors. The Long-tailed Vole (p. 182) and the Meadow Vole (p. 180) have longer, bicolored tails.

Northern Bog Lemming

Synaptomys borealis

Total Length: 3½–6 in (8.9–15 cm)

Tail Length: ½–1⅛ in (1.3–2.9 cm)

Weight: ½–1⅛ oz (14–32 g)

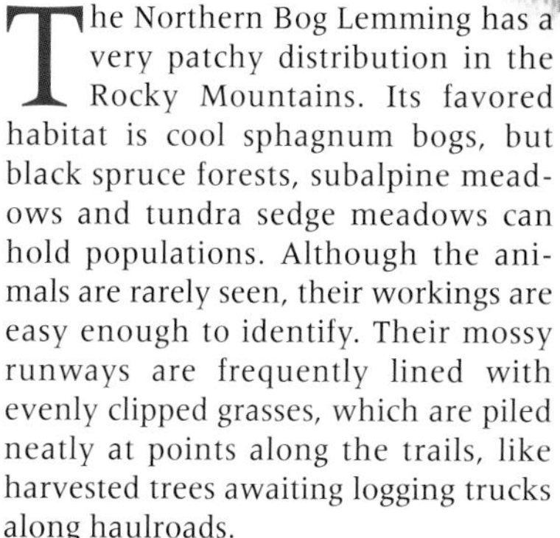

The Northern Bog Lemming has a very patchy distribution in the Rocky Mountains. Its favored habitat is cool sphagnum bogs, but black spruce forests, subalpine meadows and tundra sedge meadows can hold populations. Although the animals are rarely seen, their workings are easy enough to identify. Their mossy runways are frequently lined with evenly clipped grasses, which are piled neatly at points along the trails, like harvested trees awaiting logging trucks along haulroads.

DESCRIPTION: The ears of this stout lemming scarcely project above the fur of the head. The whole body is covered in thick fur. Although there are various color phases, the sides and back are usually chestnut or dark brown, and the underparts are usually grayish. There is a little patch of tawny-colored hairs just behind the ears. The claws are strong and curved, and the ones on the middle two front toes become greatly enlarged in winter to aid digging in frozen conditions.

HABITAT: This lemming thrives in tundra bogs, alpine meadows and even spruce woods.

FOOD: The diet is primarily composed of grasses, sedges and similar plants. If this vegetation is scarce, other emergent plants are eaten.

DEN: In summer, the nests are located about 6 in (15 cm) underground in tunnels. The nests are constructed with dry grass and fur, and there are nearby chambers for wastes. In winter, the nests are located aboveground, under the snow.

YOUNG: Little is known about the Northern Bog Lemming, but it probably breeds between spring and fall, with a gestation period of about three weeks. The litter contains two to six helpless young. Growth is rapid: they are furred by one week, weaned by three weeks, and leave to start their own families soon thereafter.

SIMILAR SPECIES: The Brown Lemming (p. 186) is generally larger and has a stubby tail and ungrooved incisors. The Long-tailed Vole (p. 182) and the Meadow Vole (p. 180) have longer tails.

RANGE: This lemming ranges across most of Alaska and Canada south of the arctic tundra. It occurs as far south as the northern parts of Washington, Idaho and western Montana.

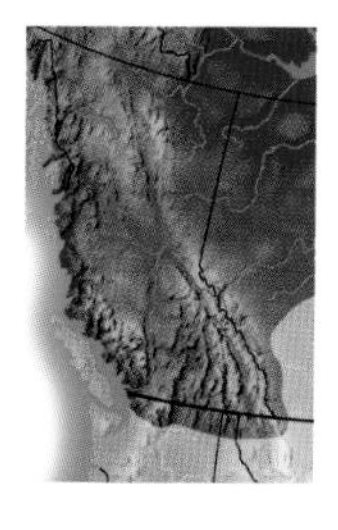

American Beaver

Castor canadensis

The American Beaver is truly a great North America mammal, and its much-valued pelt motivated the earliest of explorers to discover the riches of the Rocky Mountains. Even today, the beaver serves as an international symbol for wild places, and, quite surprisingly to many people, foreign tourists often hold out great hopes of seeing these aquatic specialists during their visits. Fortunately, the American Beaver can be regularly encountered in wet areas of the Rocky Mountains (outside of winter), where its engineering marvels can be studied in awe-inspiring detail.

Being one of the few mammals that alters its environment to suit its needs, the American Beaver often sets back natural succession and promotes changes in vegetation and animal life. Nothing seems to bother a beaver like the sound of running water, and this busy rodent builds dams of branches, mud and vegetation to slow the flow of water. The deep pools that the beaver's dams create allow it to remain active beneath the ice in winter. Such achievements require vast amounts of labor, and a single beaver may cut down as many as 1700 trees each year to ensure its survival.

Beavers live in colonies that generally consist of a pair of mated adults, their yearlings and a litter of young kits. This family group generally occupies a tightly monitored habitat that consists of several dams, terrestrial runways and a lodge. Throughout much of the Rocky Mountains, the lodge is ingeniously built of branches and mud. Adult male beavers tunnel into the banks of rivers, lakes or ponds for their den sites, and, where trees do not commonly grow or currents are swift, females may also occupy bank dens.

Although the American Beaver is not a fast mover, it more than compensates with its immense strength. It is not unusual for this firmly built rodent to handle and drag—with its jaws—a 20-lb (9.1-kg) piece of wood. The beaver's flat, scaly tail, for which it is so well known, increases an animal's stability when it is cutting a tree, and it is slapped on the water or ground to communicate alarm.

Beavers are well adapted to their aquatic lifestyles. They have valves that allow them to close their ears and nostrils when they are submerged, and clear membranes slide over the eyes. Because the lips form a seal behind the incisors, beavers can chew while they are submerged without having water,

RANGE: Beavers can be found from the northern limit of deciduous trees south to northern Mexico. They are absent only from the Great Basin, the Southwestern deserts and extensive prairie areas devoid of trees.

DID YOU KNOW?

Beavers are not bothered by lice or ticks, but there is a tiny, flat beetle that lives only in a beaver's fur and nowhere else. This beetle feeds on beaver dandruff, and its meanderings probably tickle sometimes, because beavers often scratch themselves when they are out of water.

Total Length: 3–4 ft (0.9–1.2 m)

Tail Length: 11–21 in (28–53 cm)

Weight: 18–99 lb (8.2–45 kg)

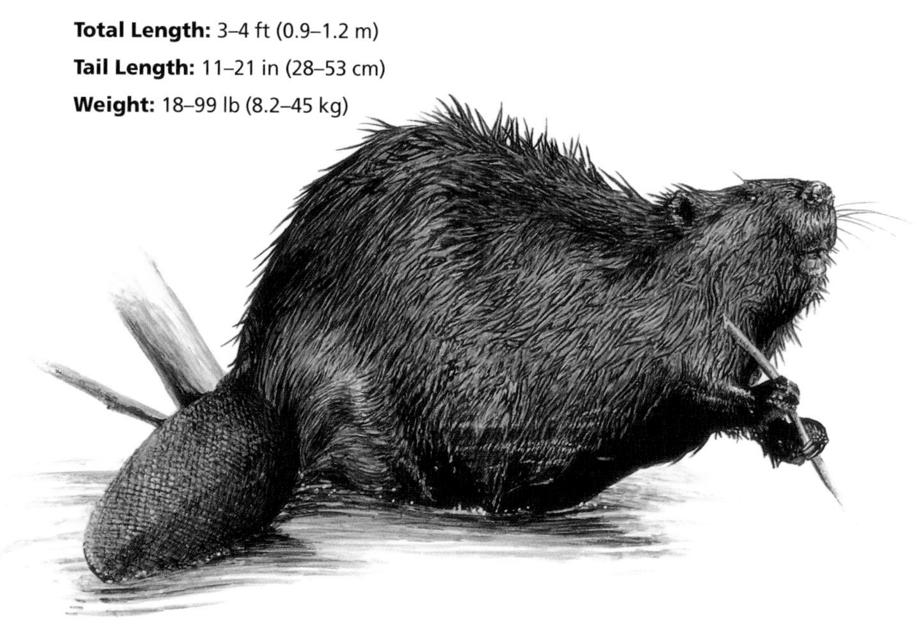

mud and chips enter the mouth. In addition to their waterproof fur, beavers have a thin layer of fat to protect them from cold waters, and the oily secretions they continually groom into their coats keeps their skin dry.

Truly, the American Beaver is an impressive animal that shapes, through its industrious workings, the physical settings of many wilderness areas. Although most tree cutting and dam building occurs at dusk or at night, you may see beavers during the day—sometimes working, but usually sunning themselves.

DESCRIPTION: This chunky, dark brown rodent has a broad, flat, scaly tail, short legs, a short neck and a broad head with short ears and massive, orange-faced incisors. The underparts are paler than the back and lack the reddish-brown hue. The nail on the next-to-outside toe of each webbed hindfoot is split horizontally, allowing it to be used as a comb in grooming the fur. The forefeet are not webbed.

HABITAT: Beavers occupy freshwater environments wherever there is suitable woody vegetation. They are sometimes even found feeding on dwarf willows above timberline.

FOOD: Bark and cambium, particularly those of aspen, willow, alder and birch, are favored, but aquatic pond vegetation is eaten in summer. Beavers sometimes come ashore to eat some grains or grasses.

DEN: Beaver lodges are cone-shaped piles of mud and sticks. Beavers construct a great mound of material first, and then they chew an underwater access tunnel into the center and hollow out a den. The lodge is typically located away from shore in still water; in flowing waters it is generally on a bank. Access to the lodge is from about 3½ ft (1.1 m) below the water's surface. A low shelf near the two or three plunge holes in the lodge allows much of the water to drain from the beavers before they enter the den chamber.

Beavers often pile more sticks and mud on the outside of the lodge each year, and shreds of bark accumulate on the den floor. Adult males generally do not live in the lodge but dig bank burrows across the water from the lodge entrance. These burrows, the entrances to which are below water, may sometimes be as long as 160 ft (50 m), but most are much shorter.

YOUNG: Most mating takes place in January or February, but occasionally as much as two months later. After a gestation period of four months, a litter of usually four kits is born. There is evidence that a second litter may be born in some years. At birth, the 12–23-oz (340–650-g) kits are fully furred, their incisors are erupted, and their eyes are nearly open. The kits begin to gnaw before they are one month old, and weaning takes place at two to three months. Beavers become sexually mature when they are about two years old, at which time they often disperse from the colony.

walking trail

SIMILAR SPECIES: The Common Muskrat (p. 184) is much smaller, and its long tail is laterally compressed rather than paddle-shaped. The Northern River Otter (p. 112) has a long, round, tapered, fur-covered tail, a streamlined body and a small head.

Common Muskrat

Olive-backed Pocket Mouse

Perognathus fasciatus

While there are some wild mammals that literally come to your backdoor, others require a visit to their special place of residence. The tiny Olive-backed Pocket Mouse is one of the latter—it is a specialized rodent that can only be found in a handful of locales in the Rockies. A staunch resident of active, open sand dunes, pocket mice are fond of dust baths: they roll and dig in the sand and then brush their fur with both their forefeet and hindlimbs. They even invert their cheek pouches to clean them against the substrate.

Like kangaroo rats, pocket mice have large hindfeet and small forelegs. They tend to sit on their hindlegs outside their burrows, but the body remains horizontal. Pocket mice move either in a slow walk or an unusual hop that involves all four limbs.

Compared to kangaroo rats, pocket mice are poor jumpers—their moderately-sized hindlegs lack the needed power for exceptionally long or high jumps. Instead, the Olive-backed Pocket Mouse typically relies on speed, agility and quick escapes into its burrow to evade such predators as hawks, owls, snakes, foxes and weasels. Individuals have been known to jump up to 2 ft (61 cm) vertically in response to a sudden alarm.

Unlike most hibernating rodents, pocket mice do not build up a store of fat; instead, they pack their burrows with vast numbers of seeds. When outside food supplies dwindle, whether because of cold winter temperatures or during periods of extreme summer heat, pocket mice retreat to their burrows and enter torpor, a state of dormancy that is not as deep as hibernation. They arouse periodically to urinate and feed on their stored seeds, but they consume less than half the amount of food eaten each day during summer. Pocket mice do not need to drink at this time of year, or at any other, because their metabolism generates water through the digestion of lipids in the seeds.

DESCRIPTION: This tiny, attractive, docile mouse has a dominant buffy back color that is modified by blackish or olive-colored hairs. The dark hairs end abruptly, and the back contrasts with the buffy sides and the white or buffy-white feet and underparts. The whole coat is shiny and soft. There is a buffy spot behind each ear. The tail is long and thin and uniformly colored. The antitragus in the base of the ear is

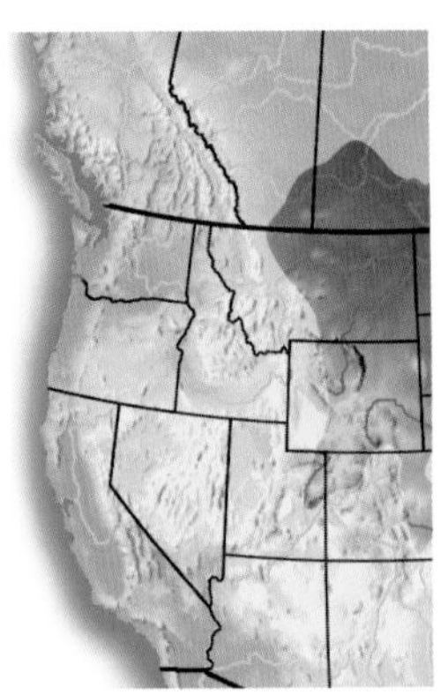

RANGE: The range of the Olive-backed Pocket Mouse extends from southeastern Alberta east to western Manitoba and south to Colorado.

DID YOU KNOW?

One study estimated that an individual mouse collects close to 300,000 tiny grass and weed seeds a year. These seed caches, used between November and March, frequently weigh more than twice as much as the collector does prior to winter.

Total Length: 4–6 in (10–15 cm)

Tail Length: 2–2¾ in (5.1–7 cm)

Weight: ¼–½ oz (7–14 g)

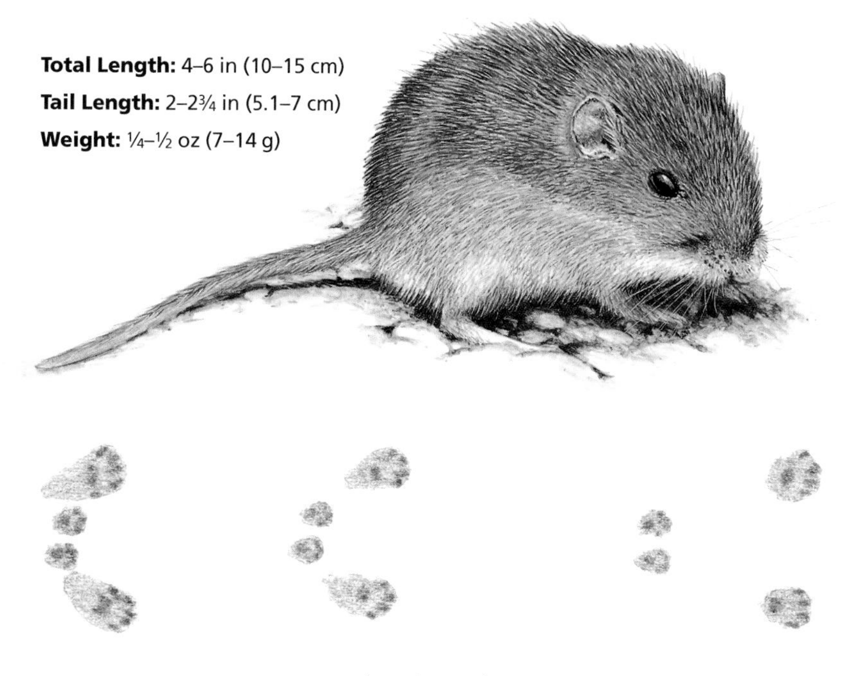

hopping trail

not lobed. The auditory bulla of all pocket mice are huge and extend onto the upper surface of the skull.

HABITAT: The Olive-backed Pocket Mouse is restricted to grasslands on light, sandy soils.

FOOD: Seeds with a higher than average oil content, including Russian thistle, knotweed, tumbling mustard, lambs-quarter, blue-eyed grass and foxtail, seem to be the mainstay of the diet. Some green vegetation and a few insects are also consumed, and grasshopper eggs may be stored in the burrow.

DEN: Summer tunnels, which are about ¾ in (1.9 cm) in diameter, form a network of storage and refuge burrows that are 1–1½ ft (30–46 cm) deep. The tunnel entrances are plugged with soil by day. The summer nesting chamber is bare. Winter burrows may penetrate 6½ ft (2 m) beneath the surface. They contain a grass nest, and the first 3½ ft (1.1 m) of the burrow is plugged with soil. The entire system may have a diameter of 33 ft (10 m).

YOUNG: Breeding begins after the pocket mice emerge in spring. After four weeks of gestation, a litter of usually four to six young is born. There are two litters each season. Young pocket mice become sexually mature the spring after their birth.

SIMILAR SPECIES: The Great Basin Pocket Mouse (p. 194) has a more western range, and a narrow buffy line separates the upperparts from the underparts. The Silky Pocket Mouse (p. 195) has a shorter tail and a more southern range. The Ord's Kangaroo Rat (p. 196) is larger, has an extremely long tail and rests with its body approaching a more upright stance.

Great Basin Pocket Mouse

Perognathus parvus

Total Length: 6–8 in (15–20 cm)

Tail Length: 3⅜–4½ in (8.6–11 cm)

Weight: ½–1 oz (14–28 g)

For Great Basin Pocket Mice, a bright future is one that is invested in seeds. This animal collects massive quantities of seeds and burrows down 3–6 ft (0.9–1.8 m) to spend the winter. The number of seeds stored by an individual is even more phenomenal when you consider that each seed is handled individually and that most are smaller than the head of a pin. Equally outstanding is the fact that in some parts of the range there is considerable competition between pocket mice and ants for the seeds.

DESCRIPTION: The back is a glossy, yellowish buff, with the many black-tipped hairs that overlay the fur giving a peppered appearance. A narrow, buffy line separates the back color from the uniform white or buffy-white underparts and feet. The tail is generally more than half the animal's total length, and it is darker above and lighter below. The hindfoot is ⅞–1 in (2.2–2.5 cm) long—either shorter or longer than those of the other Rocky Mountain species.

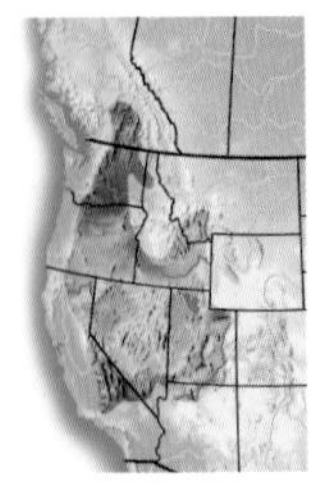

RANGE: The Great Basin Pocket Mouse is found in southwestern Wyoming and in Utah and Nevada. It also occupies semi-desert areas in Idaho, Washington, Oregon, California and Arizona.

HABITAT: These pocket mice live in sandy soils in arid areas. Most inhabit the plains, but a few are found in the foothills west of the continental divide.

FOOD: Cheatgrass seeds form the bulk of the diet, but this mouse also eats the grain and seedlings of winter wheat, and, in spring, considerable numbers of insects. Caterpillars and adult insects supplement a diet that includes the seeds of Russian thistle, wild mustards, bitterbrush and pigweed.

DEN: The extensive burrow system resembles that of the Olive-backed Pocket Mouse, with shallow summer tunnels and deep winter burrows. The tunnel entrances are plugged with soil by day and throughout winter. The summer nesting chamber is bare, but a grass nest is built for winter.

YOUNG: A female mates soon after she emerges from hibernation in April. About four to six helpless young are born after a gestation period of 21 to 28 days. The young are weaned at 25 days, when they weigh about ¼ oz (7.1 g). A few females have a second litter in late summer.

SIMILAR SPECIES: Neither the Silky Pocket Mouse (p. 195) nor the Olive-backed Pocket Mouse (p. 192) have the lobed antitragus seen in the ear of the Great Basin Pocket Mouse.

Silky Pocket Mouse

Perognathus flavus

Total Length: 4–4¾ in (10–12 cm)

Tail Length: 1¾–2⅜ in (4.5–6 cm)

Weight: 3/16–⅜ oz (5.3–11 g)

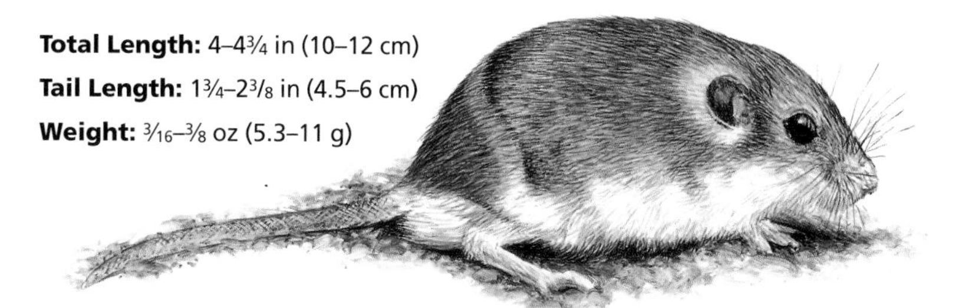

As if part of an intricately plotted spy movie, the Silky Pocket Mouse builds hidden doorways into and out of its underworld dens. Its burrows have side tunnels that end just below the surface, and the pocket mouse can break in through the little cap of soil to escape the pursuit of a snake, weasel, skunk or badger. Once tucked underground in its multi-chambered, shallow refuge, a Silky Pocket Mouse is safe from just about any marauder. Should the threat find its way into the burrow, however, the pocket mouse uses the reverse trick and exits through a seeming dead end.

DESCRIPTION: This pocket mouse is the smallest of the three found in the Rockies, and its tail is less than half its total length. The soft fur of the back is pinkish buff, with a fine mixture of blackish hairs. There is a poorly defined buffy wash between the color of the back and the white to whitish-buff belly. The clear buffy spot behind each ear is larger than the ear.

HABITAT: Arid, sandy plains are the preferred habitat, although individuals have been trapped at up to 8000 ft (2440 m) along the east slopes of the Colorado Rockies.

FOOD: The diet consists of the seeds of grasses and forbs and occasionally vegetation and insects.

DEN: The extensive burrow system resembles that of the Olive-backed Pocket Mouse, with shallow summer tunnels and deep winter burrows. The tunnel entrances are plugged with soil by day and throughout winter. A male usually maintains a greater number of burrow systems than a female—it is not unusual for a male to have six or seven separate burrow sites.

YOUNG: Breeding occurs after the emergence of females from hibernation, usually in late March. There are usually three to six young in a litter. Information on their development is lacking, but there seem to be two litters each summer, and those born in one summer are sexually mature the next spring.

SIMILAR SPECIES: The Olive-backed Pocket Mouse (p. 192) has a longer tail and a more distinct buffy side stripe. The Great Basin Pocket Mouse (p. 195) has a lobed antitragus in the ear.

RANGE: The Silky Pocket Mouse has been found from southeastern Wyoming and western Nebraska south to Mexico. The westernmost populations occur in northern and southern Arizona.

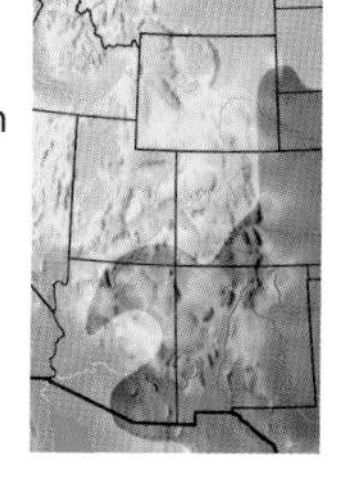

Ord's Kangaroo Rat

Dipodomys ordii

Finding sand dunes within this gentle rodent's range is challenging in itself; locating an individual kangaroo rat requires luck and knowledge. This nocturnal hopper is best seen on dark, moonless and overcast nights. If you shine a flashlight across the dunes or drive slowly along sandy roadcuts, this sand-dweller might be revealed. By day, kangaroo rats retire to their plugged sand burrows.

Much of a kangaroo rat's food is taken from the sand; it forages slowly, sifting out seeds with its sharp foreclaws. Seeds that are to be eaten immediately are first husked, but those to be stored are left intact. The kangaroo rat transports the food to its burrow in its spacious external cheek pouches, which can be turned inside-out for cleaning and combing with the foreclaws. Sand is critical to this mammal's cleanliness, and dust bathing is an important part of a kangaroo rat's grooming routine.

The Ord's Kangaroo Rat can live its entire life without drinking free water. It can survive on the metabolic water produced through the breakdown of the oils and fats in the seeds it eats. This water is used in all body functions, including digestion, excretion, reproduction and milk production. Despite this ability, in the wild, kangaroo rats will lap droplets of dew and sometimes eat a bit of green vegetation when they are available.

DESCRIPTION: The back is yellowish buff, with a few black hairs down the center, the sides are clear buff, and the belly is white. The eyes are large, luminous and protruding. There is a white spot above the eye and behind the brownish-black ear. A black patch on the side of the nose, above the white lip, marks the base of the whiskers. There is a diagonal, white line across the hip. The extremely long hindfeet are white on top and brownish black on their hairy soles. The greatly reduced forelegs are held up and are often not visible in profile as the animal sits hunched over, supporting itself on its hindfeet. The tail, which is at least as long as the body, is tufted at the tip. It has white sides and brownish-black upper and lower surfaces.

HABITAT: This kangaroo rat occupies sandy, grassland and sagebrush semi-desert sites. Disturbed, vegetation-free areas, whether produced by drifting sand or by road building or traffic, seem particularly attractive.

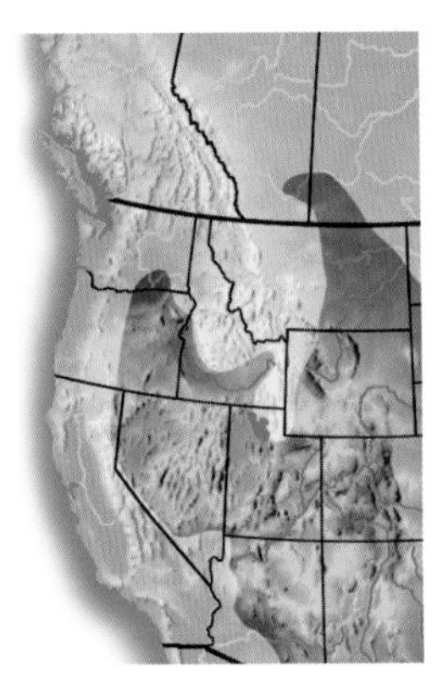

RANGE: This widely distributed kangaroo rat occurs from southeastern Alberta and southwestern Saskatchewan south through the Great Plains and western Texas into Mexico, and from eastern Oregon south through the Great Basin and Arizona.

DID YOU KNOW?

The Ord's Kangaroo Rat's huge auditory bullae make up a major rear portion of the skull. Able to hear very low-frequency sounds, kangaroo rats can avoid a rattlesnake's bite, presumably because they are able to hear the snake's movement.

Total Length: 9–11 in (23–28 cm)

Tail Length: 5½–6¼ in (14–16 cm)

Weight: 1½–3⅜ oz (43–96 g)

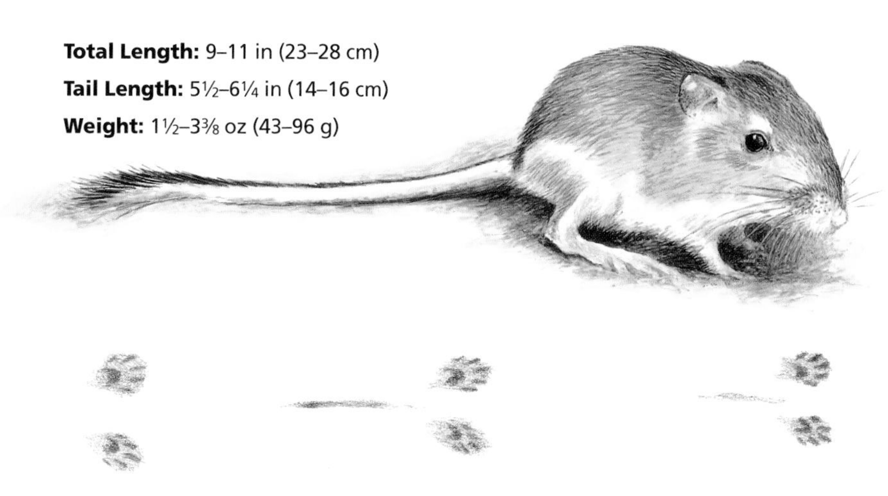

hopping trail

FOOD: Seeds make up more than three-quarters of the year-round diet. Insects, such as ants, butterfly pupae, adult beetles and larval antlions, account for one-fifth of the diet in spring, and grasshoppers and roots are eaten in summer.

DEN: Burrows, which are usually located in the sides of sand dunes, dry eroded channels or road slopes, are about 3 in (7.6 cm) in diameter. The entrances are plugged during the day. The tunnels branch frequently, with some branches used for food storage and at least one as a nesting chamber. Most of the burrow system is often within 1 ft (30 cm) of the surface. The burrow is actively defended against other kangaroo rats.

YOUNG: Breeding occurs in early to mid-spring, and sometimes again in mid-summer. After a 29- to 30-day gestation period, a litter of usually three to five young is born in a nest built just beforehand. The helpless newborns, which weigh about 3/16 oz (5.3 g), are groomed by the mother. At two weeks their eyes open, and at three weeks they have functioning cheek pouches. After the young reach adult size, at five to six weeks, they disperse to develop their own burrows. The mortality rate in the first year may be as high as 80 percent.

SIMILAR SPECIES: No other rodent has the extremely long tail and hindlegs, big head and stocky body of the Ord's Kangaroo Rat. The Olive-backed Pocket Mouse (p. 192) is smaller, has a shorter tail and usually holds its body more horizontally when it rests on its hindfeet.

hopping group

Northern Pocket Gopher

Thomomys talpoides

The Northern Pocket Gopher is one of nature's rototillers. This ground-dwelling rodent is continuously tunneling through dark, rich soils, and one individual is capable of turning over 16 tons of soil every year. Evidence of its workings is commonplace on the land, because the freshly churned earth is neatly piled in mounds, or "gopher cores," without visible entrances. In many agricultural areas, this mammal's mounds, which can damage machinery and cover vegetation, make it the most controlled nuisance mammal. Many people incorrectly call pocket gophers "moles," but there are no true moles in the Rocky Mountains.

Pocket gopher push-ups hide the access holes to a system of burrows. From the rodent's viewpoint, the surface provides a space to dump the dirt from tunnel excavation. When the ground is covered by snow, pocket gophers still bring up waste soil to the surface and pack it into snow tunnels. When the snow melts, these soil cores, or "crotovinas," are left exposed.

There is no mammal in the Rockies better adapted to an underground existence. The Northern Pocket Gopher has small eyes, which it rarely needs in its darkened world; reduced external ears that do not interfere with tunneling; short, lax fur that does not impede backward or forward movement in the tunnels; and a short, sparsely haired tail that serves as a tactile organ when the animal is tunnel-running in reverse.

To dig its elaborate burrows, the Northern Pocket Gopher has heavy, stout claws on short, strong forelegs and a massive lower jaw armed with long incisors. Once the soil is loosened with tooth and claw, it is pushed back under the body, initially with the forefeet, and then further with the hindfeet. When sufficient soil has accumulated behind the animal, the gopher turns, guides the mound with its forefeet and head and pushes with its hindlegs until the soil is in a side tunnel or on the surface.

Pocket gophers are named for their large, externally opening, fur-lined cheek pouches. As in the related pocket mice and kangaroo rats, these "pockets" are used to transport food, but they have no direct opening to the animal's mouth.

DESCRIPTION: This squat, bullet-headed rodent has visible incisors, long foreclaws and a thick, nearly hairless tail. A row of stiff hairs surrounds the naked soles of the forefeet. The upperparts, which are slightly darker than the

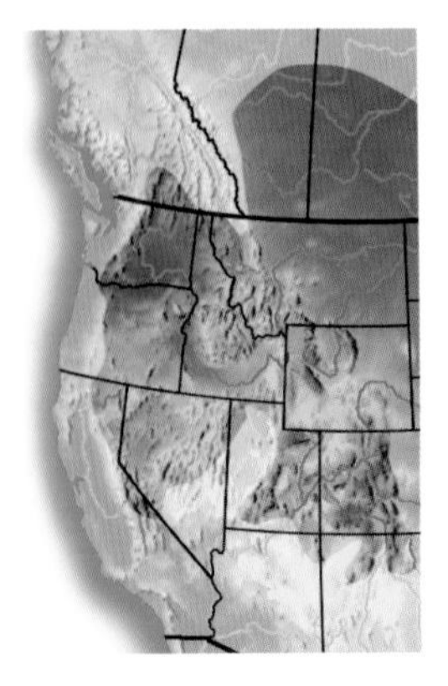

RANGE: This species occupies most of the northern Great Plains and western mountains from Manitoba to British Columbia and south to Nebraska, Colorado, northern Arizona and New Mexico. It occurs as far east as western Minnesota, and as far west as western Washington and Oregon.

DID YOU KNOW?

A pocket gopher's incisor teeth can grow at a spectacular rate: lower incisors are reported to grow as much as 0.04 in (1 mm) a day; upper incisors grow 0.02 in (0.5 mm) a day. If that rate was continuous, the lower incisors could grow 14 in (36 cm) in a year.

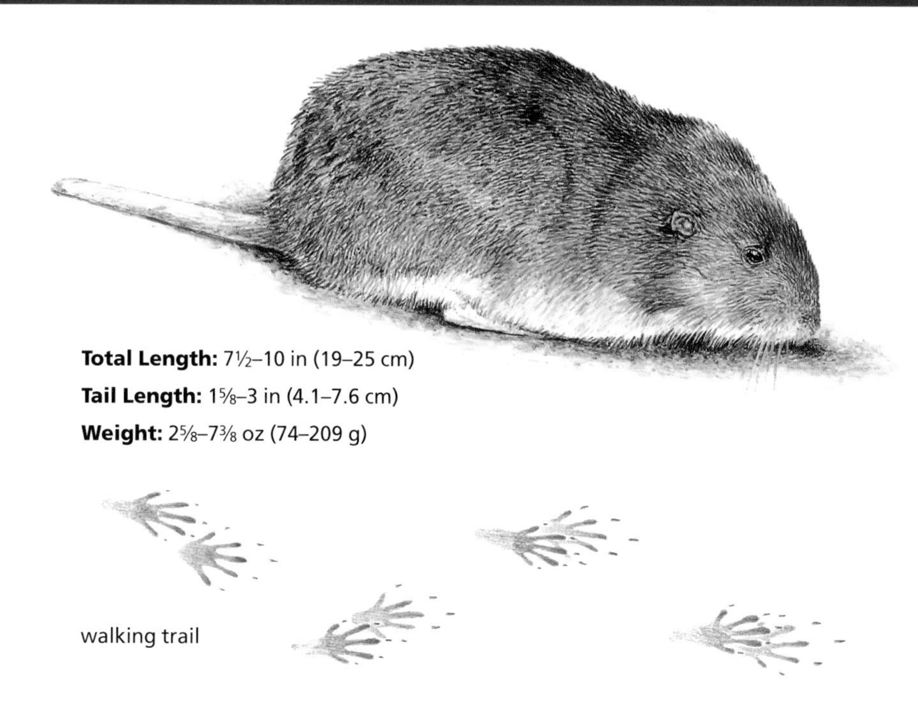

Total Length: 7½–10 in (19–25 cm)

Tail Length: 1⅝–3 in (4.1–7.6 cm)

Weight: 2⅝–7⅜ oz (74–209 g)

underparts, often match the soil color—individuals may be black, dark gray, brown or even light gray.

HABITAT: This adaptable animal avoids only dense forests, wet or waterlogged, fine-textured soils, very shallow rocky soils or areas exposed to strong winter freezing of the soil.

FOOD: Succulent underground plant parts are the staple diet, but in summer pocket gophers leave their burrows at night to collect green vegetation.

DEN: The burrow system may spread 150–500 ft (46–150 m) laterally and extend 2 in (5.1 cm) to 10 ft (3 m) deep. The tunnels are about 2 in (5.1 cm) in diameter. Some lateral tunnels serve for food storage, others as latrines. Several nesting chambers, 8–10 in (20–25 cm) in diameter and filled with fine grass, are located below the frost line. Spoil from tunneling is spread fanwise to one side of the burrow entrance; then the burrow is plugged from below. Only a single gopher occupies a burrow system, except during the breeding season, when a male may share a female's burrow for a time.

YOUNG: Breeding occurs once a year, in April or May. Following a 19- to 20-day gestation, three to six young are born in a grass-lined nest. Weaning takes place at about 40 days. When the young weigh about 1½ oz (42 g) they leave to either occupy a vacant burrow system or begin digging their own. They are sexually mature the following spring.

SIMILAR SPECIES: The Idaho Pocket Gopher (p. 200) is generally smaller and has a very restricted range. The Botta's Pocket Gopher (p. 201) looks very similar. Voles (pp. 176–83) do not have the large external cheek pouches, nearly hairless tail or long front claws of pocket gophers, although their color patterns are similar.

Idaho Pocket Gopher

Thomomys idahoensis

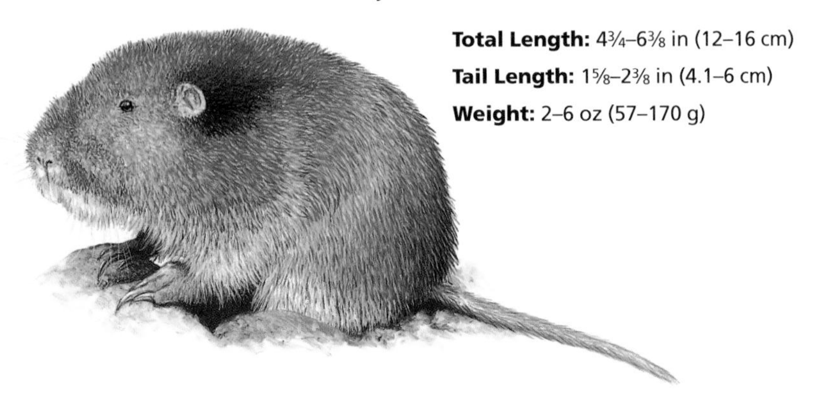

Total Length: 4¾–6⅜ in (12–16 cm)

Tail Length: 1⅝–2⅜ in (4.1–6 cm)

Weight: 2–6 oz (57–170 g)

The incisors of an Idaho Pocket Gopher remain outside the lips even when the mouth is closed. This characteristic allows the gopher to eat food underground or dig through the earth without getting soil particles in its mouth. Another unusual characteristic of pocket gophers is the external cheek pouches that are used to carry food and nesting material. Daily grooming involves cleaning these fur-lined pouches. The cheek pouches can be turned inside-out to make grooming them easier. Special muscles in the cheek maintain the pouches and pull them back inside after the grooming is finished.

DESCRIPTION: These gophers are quite variable in color, ranging from yellowish brown to sooty gray. They are darker on their backs than on their undersides. Their tails are lightly furred and are grayish in color. Their ears are rounded.

RANGE: The Idaho Pocket Gopher is found only in extreme southwestern Montana, southeastern Idaho and northern Utah.

HABITAT: Idaho Pocket Gophers live in areas of deep soil, such as river valleys and old lake beds, in dry, treeless ranges.

DEN: As with other pocket gophers, this animal spends most of its life underground. It sleeps in special nest chambers, and has separate tunnels for wastes. Deep lateral tunnels are for dens and food storage, while shallow tunnels are foraging routes.

FOOD: These pocket gophers feed on vegetation of all sorts, especially roots, tubers and shoots protruding into their burrows. They also consume some aerial plant parts, such as leaves, seeds and fruit.

YOUNG: Idaho Pocket Gophers may have two or more litters in one season if conditions are good. Litter sizes range from three to seven young, and gestation is about 18 days.

SIMILAR SPECIES: All other pocket gophers are larger and have much broader ranges. The Botta's Pocket Gopher (p. 201) is usually browner.

Botta's Pocket Gopher

Thomomys bottae

Total Length: 6⅝–11 in (17–28 cm)

Tail Length: 1⅝–3¾ in (4.1–9.5 cm)

Weight: 2½–8⅞ oz (71–250 g)

People have mixed feelings about pocket gophers. In natural areas, they are an important component of the ecosystem. Annually, these animals turn up large volumes of soil, which aerates the ground, cycles the soil nutrients and aids water absorption. Studies have shown that where pocket gophers live in normal numbers, plants grow better. Gopher mounds can interfere with agricultural machinery, however, and the gophers may compete with livestock for vegetation.

DESCRIPTION: Botta's Pocket Gophers are dark or grayish brown above and slightly paler below. Their tails are sparsely furred and are tawny or gray in color. Their ears are rounded and inconspicuous.

HABITAT: These pocket gophers live in a variety of habitats and soil types, from deserts to mountain meadows and from sandy to clay soils.

DEN: Like other pocket gophers, the Botta's spends most of its life underground. It digs nest chambers, special wastes tunnels, deep lateral tunnels for dens and food storage and shallow tunnels for foraging routes.

FOOD: Pocket gophers feed on vegetation of all sorts, especially roots and tubers they encounter while burrowing and shoots they pull down into their burrows. They also consume some aerial plant parts, such as leaves, seeds and fruit.

YOUNG: Botta's Pocket Gophers may have several litters in a season. In ideal habitats, they may breed throughout the year. Litters average six young each, and the gestation period is 18 or 19 days.

SIMILAR SPECIES: The Northern Pocket Gopher (p. 198) is generally not as brown, and where both species occur, it is found in drier, stony upland sites. The Idaho Pocket Gopher (p. 200) is usually grayer and occurs in a smaller, more northern range.

RANGE: The Botta's Pocket Gopher is found from southwestern Oregon through California and southeast to Colorado and western Texas.

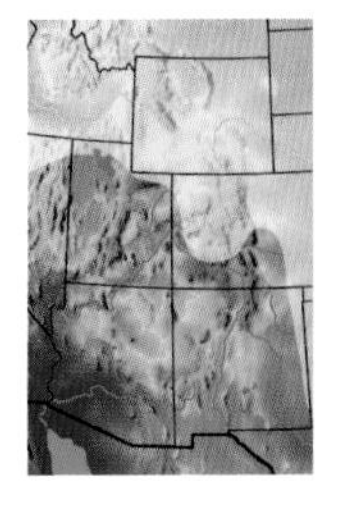

Least Chipmunk

Tamias minimus

The sound of scurrying among fallen leaves, a flash of movement and sharp, high-pitched "chips" are often enough to direct your attention to the nervous behavior of a Least Chipmunk. Using fallen logs as runways and the leaf litter as its pantry, this busy animal inhabits wooded areas throughout much of the Rocky Mountains. With its extensive range, it is one of the most commonly seen chipmunks in parts of the Rockies.

The word "chipmunk" is thought to be derived from the Algonkian word for "head first," which is the manner in which a chipmunk descends a tree, but, contrary to cartoon-inspired Disney myths, chipmunks spend very little time in high trees. They prefer the ground, where they bury food and dig golf ball–sized entrances to their networks of tunnels. Chipmunk burrows are known for their well-hidden entrances, which never have piles of dirt to give away their locations.

In certain heavily visited parks and golf courses, Least Chipmunks that have grown accustomed to human handouts can be very easy to approach. These exchanges contrast dramatically with the typically brief sightings of wild chipmunks, which scamper away at the first sight of humans. In the wild, chipmunks rely on their nervous instincts to survive in their predator-filled world.

DESCRIPTION: This tiny chipmunk has three dark and two light stripes on its face, and five dark and four light stripes on its body. The central dark stripe runs from the head to the base of the tail, but the other dark stripes end at the hips. The overall color is grayer and paler than other chipmunks, and the underside of the tail is yellower. The tail is quite long—more than 40 percent of the total length.

HABITAT: The Least Chipmunk inhabits a wide variety of areas, including open coniferous forests, sagebrush flats, rocky outcroppings and pastures with small shrubs. It may be seen at ranches or farms well away from mountains or forests, attracted there by livestock feed.

FOOD: This chipmunk loves to dine on ripe berries, such as chokecherries, pincherries, strawberries, raspberries or blueberries. Other staples in the diet include nuts, seeds, grasses, mushrooms and even insects and some other animals. It may be an important predator

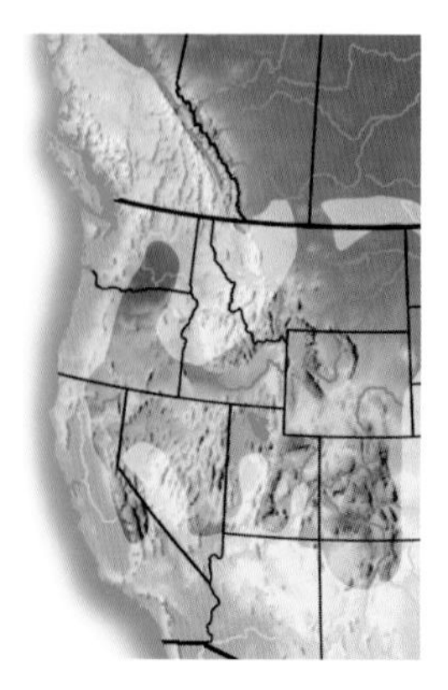

RANGE: The extensive range of this species spreads from the central Yukon to western Quebec, and in the U.S from Washington to northern California, from North Dakota to New Mexico, and east to just west of the Great Lakes.

DID YOU KNOW?

In summer, a chipmunk's body temperature is 95°–108° F (35–42° C). In winter, when it is hibernating in its burrow, its body temperature drops to 41°–45° F (5°–7° C).

Total Length: 7–9½ in (18–24 cm)
Tail Length: 3–4¼ in (7.6–11 cm)
Weight: 1¼–2½ oz (35–71 g)

bounding trail

on eggs and nestling birds during the nesting season. A chipmunk may be attracted to animal feed around mountain ranches, and sometimes one will be filling its cheek pouches from a pile oats shared by a horse.

DEN: The majority of Least Chipmunks den in underground burrows, which have concealed entrances, but some individuals live in tree cavities or even make spherical leaf and twigs nests among the branches in the manner of tree squirrels.

YOUNG: Breeding occurs about two weeks after the chipmunks emerge from hibernation in spring. After about a one-month gestation, a litter of two to seven (usually four to six) helpless young is born in a grass-lined nest chamber. The young develop rapidly, and the mother may later transfer them to a tree cavity or tree nest.

SIMILAR SPECIES: It is very difficult to distinguish chipmunk species in the field, but range maps often help. The Yellow-pine Chipmunk (p. 204) tends to have brighter colors than the Least Chipmunk. The Red-tailed Chipmunk (p. 209) has a brick red tail underside. The larger Golden-mantled Ground Squirrel (p. 230), with its bold side stripes, is often mistaken for a chipmunk, but only chipmunks have the horizontal face stripes. The Golden-mantled Ground Squirrel has no facial stripes.

foreprint

hindprint

Yellow-pine Chipmunk

Tamias amoenus

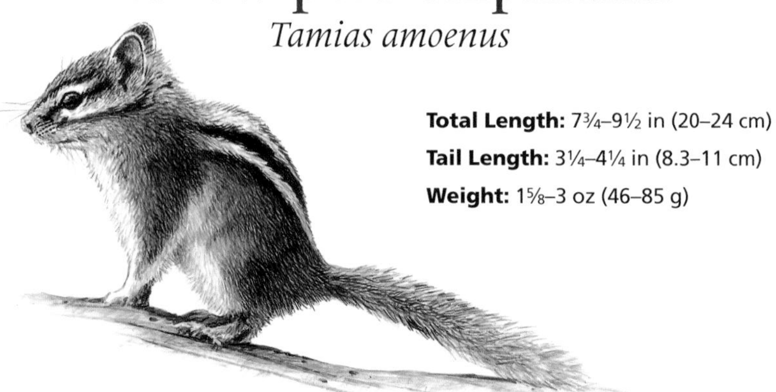

Total Length: 7¾–9½ in (20–24 cm)

Tail Length: 3¼–4¼ in (8.3–11 cm)

Weight: 1⅝–3 oz (46–85 g)

Perhaps more common in the mountains than the wide-ranging Least Chipmunk, the Yellow-pine Chipmunk can be seen by anyone who is willing to invest the time and effort in a search. It can often be found near semi-open day-use areas in mountain parks, although it tends not to be among such front-line rodents as the Golden-mantled Ground Squirrel and the Columbian Ground Squirrel, which learn to beg for attention and handouts. The Yellow-pine Chipmunk neither seeks nor completely shuns the company of curious humans.

DESCRIPTION: This brightly colored chipmunk is tawny to pinkish cinnamon. There are three dark and two light stripes on the face, and five dark and four light stripes on the back. The light stripes are white or grayish. The dark stripes are nearly black, and the central three extend all the way to the rump. The sides of the body and the underside of the tail are grayish yellow. A female tends to be larger than a male.

RANGE: This mountain chipmunk occurs in British Columbia, extreme western Alberta and the northwestern U.S.

HABITAT: This chipmunk is common in brushy or rocky areas of coniferous mountain forests.

FOOD: The bulk of the diet is composed of conifer seeds, nuts, some berries and insects. It is common for chipmunks to eat eggs, fledgling birds, young mice or even carrion.

DEN: The Yellow-pine Chipmunk usually lives in a burrow that has a concealed entrance. It can sometimes be found in a tree cavity, but it seldom builds a tree nest.

YOUNG: The young are born in May or June, after about one month of gestation. Usually five or six young are born in a grass-lined chamber in the burrow. They are blind and hairless at birth, but their growth is rapid, and they are usually weaned in about six weeks.

SIMILAR SPECIES: The smaller Least Chipmunk (p. 202) has duller colors. The Red-tailed Chipmunk (p. 209) is larger and may be grayer, but it is most easily distinguished by the reddish underside of its tail.

Cliff Chipmunk

Tamias dorsalis

Total Length: 7⅝–11 in (19–28 cm)

Tail Length: 3⅜–5½ in (8.6–14 cm)

Weight: 2–3 oz (57–85 g)

As its name suggests, the Cliff Chipmunk makes its home in rocky areas. It is the most heat tolerant of all the mountain chipmunks, occupying microclimates that other chipmunks avoid, and it occurs only in the southernmost part of the Rocky Mountains. There is a large range of elevations where one can expect to find the Cliff Chipmunk, but this relates to exposure and not to any mountaineering requirements by the animal. On south-facing slopes that receive more sunlight and heat, Cliff Chipmunks have been found at up to 11,000 ft (3350 m). Conversely, on north-facing slopes, the cool coniferous forests that dominate at high elevations discourage Cliff Chipmunk occupation.

DESCRIPTION: This rather large chipmunk is generally grayish brown, with rather indistinct side stripes. Its facial stripes are clearer than the stripes over the back. The belly is dull white, tinged at times with buff. The long tail is bushy, with slight red highlights.

HABITAT: Cliff Chipmunks are most frequently found in rocky areas and near cliffs in the pinyon-juniper zone of the southern Rocky Mountains.

FOOD: Juniper berries are reported to be the most common food item, but the seeds of pinyon pine, yellow pine, serviceberry, Russian thistle and sagebrush are also eaten.

DEN: The Cliff Chipmunk builds its nest in a burrow or uncommonly in a hollow log. The nest chamber, which is about 7 in (18 cm) in diameter, is often positioned within the roots of a shrub, and it is lined with dried grass and shredded bark.

YOUNG: These chipmunks mate in April and May. Some females may have more than one litter a year if they mate again in late summer. The average litter size is four to eight.

SIMILAR SPECIES: Identifying chipmunk species can be difficult. Among the Rocky Mountain species, the Cliff Chipmunk has the least distinct, grayest body stripes.

RANGE: This chipmunk ranges south from southeastern Idaho and southwestern Wyoming through western Nevada, Utah, Arizona and eastern New Mexico.

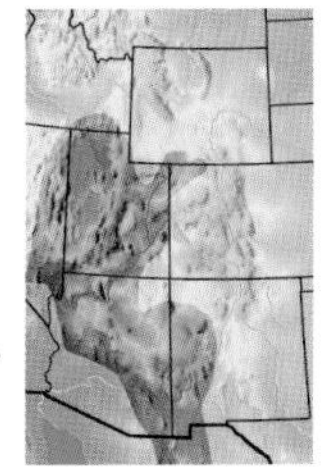

Colorado Chipmunk

Tamias quadrivittatus

Wherever you travel along the spine of the Rocky Mountains, you are not far from a chipmunk. These little striped squirrels occupy a great diversity of habitats in the Rockies—they can be seen from the loftiest alpine areas down to the arid grasslands—although they typically inhabit areas with at least some forest cover.

Certain individual chipmunks become exceptional in their acclimatization to humans. Their overt curiosity, combined with our ill-advised but well-intentioned handouts, produces some chipmunks that are so accustomed to people that they have been known to climb up the legs of someone offering food. Although generally harmless, this overzealous friendliness is understandably alarming to people unaccustomed to it. Of course, anyone reacting violently to the perceived attack of one of these minuscule squirrels becomes an instant target of laughter and the subject of stories for years.

The Colorado Chipmunk, like most of its relatives, is often seen scampering across open areas or roads with its tail held vertically. Sometimes it is heard prior to being seen. Its high-pitched "chips," given while the chipmunk wags its tail from side to side, sound similar to the notes of sparrows and other songbirds. Once the tone is learned, however, it is difficult to walk far through a western forest without knowingly being in the company of this rodent or its kin.

The Colorado Chipmunk is easily mistaken for one of the many other species of chipmunks found in the southern and central U.S. Rockies. The Colorado Chipmunk is generally the most showy of this select group, however, and it has a brownish rather than gray neck.

Like other chipmunks, Colorado Chipmunks are sometimes predators of the eggs and nestlings of small birds. Nests in shrubs are especially vulnerable, but chipmunks may also climb high into trees to obtain these protein-rich foods.

DESCRIPTION: Chipmunks have a pattern of alternating dark and light stripes that extend from the tip of the nose to the rump. Only chipmunks have face stripes combined with a dark median dorsal stripe with two parallel dark stripes on each side. The background color is yellowish gray to chestnut on the sides and lighter below. The tail is moderately bushy and is often carried erect as the animal runs.

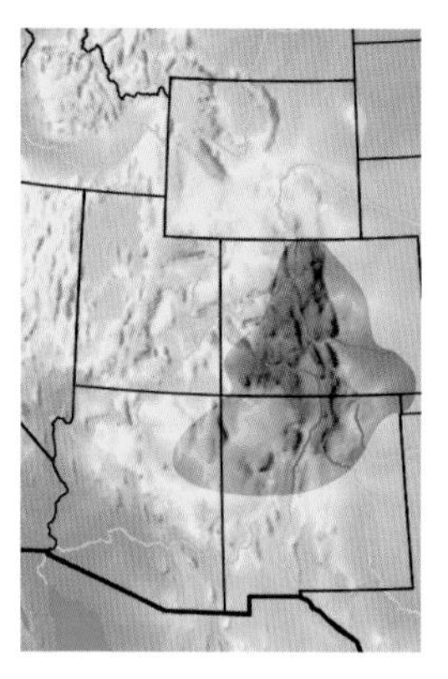

RANGE: The Colorado Chipmunk is found in Colorado, eastern Utah, northern New Mexico and the extreme northeastern corner of Arizona.

DID YOU KNOW?

This chipmunk's sharp claws and dauntless nature give it access to nearly any food source. It is a connoisseur of spruce cones and will fearlessly climb to the tops of spruce trees to dine on these choice seeds.

Total Length: 8¼–9¼ in (21–24 cm)

Tail Length: 3⅜–4⅛ in (8.6–10 cm)

Weight: about 2 oz (57 g)

HABITAT: The Colorado Chipmunk occupies pinyon-juniper and spruce-fir forests and open, rocky, brushy areas. It is not seen above timberline.

FOOD: Most of the diet is vegetation, consisting of pinyon nuts, acorns, spruce seeds, fungi and insects, but chipmunks also eat bird eggs and nestlings.

DEN: The Colorado Chipmunk builds its nest in a burrow or uncommonly in a hollow log. The nest chamber is about 7 in (18 cm) in diameter, is often positioned within the roots of a shrub and is lined with dried grass and shredded bark.

YOUNG: Mating occurs in spring, and after a gestation period of about one month, two to six young are born. Within the Rockies, there is likely only one litter a year.

SIMILAR SPECIES: The Uinta Chipmunk (p. 210) has a yellowish tinge on its head. The Least Chipmunk (p. 202) flicks its tail up and down, not side to side, when it calls. Often, habitat and geographic location are the best clues to narrow your choice of possible species. If you see a single chipmunk within the geographic range of two or more species, you may have to resign yourself to not being able to identify it precisely. Experts often depend upon slight differences in size, coloration, behavior and the shape of the male's penis bone to distinguish species.

foreprint

Hopi Chipmunk

Tamias rufus

Total Length: 6⅝–9 in (17–23 cm)

Tail Length: 3–4¼ in (7.6–11 cm)

Weight: 1½–2 oz (43–57 g)

The constant hoarding of food by Hopi Chipmunks and their kin is not without explanation. Chipmunks do not enter a long-term dormancy and live off body fat like marmots do; instead, they have short periods of torpor interspersed with wakeful periods during which they feed and eliminate waste. Depending on the latitude and climate, a chipmunk's dormant periods may last from a few days to a couple of weeks. Stored food is vital to a chipmunk's survival in winter—if a chipmunk were to wake and find its caches empty, it would be facing starvation. Hopi Chipmunks spend at least from November to late February in their winter dens.

DESCRIPTION: The Hopi Chipmunk has a distinct orangish coloration. The dark stripes are auburn in color, rather than black, and the outermost dark stripes may be quite diffuse or even nonexistent. The tail is chestnut colored, with flecks of black above. Behind the ears there is a small but distinct patch of whitish fur.

HABITAT: This chipmunk is found in canyonlands with rocky outcroppings and areas of sparse vegetation. Associated plants include juniper, pinyon pine and hardy evergreen shrubs.

FOOD: Although Hopi Chipmunks feed on grasses, conifers seeds and green vegetation, they have a particular fondness for the cones of one-seeded juniper. They may also eat some animal matter.

DEN: This chipmunk builds its nest primarily in rock crevices, but sometimes in a hollow log. The nest chamber is a little larger than a softball and is lined with dried grass and shredded bark.

YOUNG: The Hopi Chipmunk mates in spring. Gestation lasts 30 to 33 days, and a litter contains two to six young. Within the Rockies there is likely only one litter a year.

SIMILAR SPECIES: The Colorado Chipmunk (p. 206) is a bit larger. The Least Chipmunk (p. 202) looks similar, but it is usually smaller, is generally duller in color and it tends to hold its tail more upright.

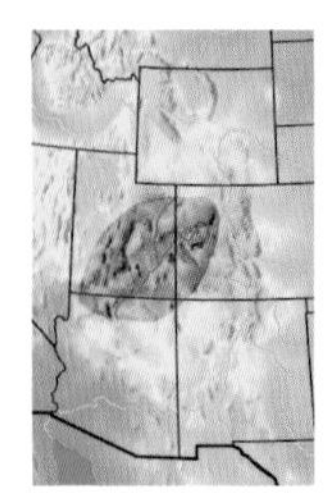

RANGE: The Hopi Chipmunk inhabits a small range in northern Arizona, western Colorado and eastern Utah.

Red-tailed Chipmunk

Tamias ruficaudus

Total Length: 8¼–9¾ in (21–25 cm)

Tail Length: 3¾–4¾ in (9.5–12 cm)

Weight: 1⅞–2⅝ oz (53–74 g)

Outside of some mountain campgrounds, Red-tailed Chipmunks require an effort to discover—they are otherwise common only in high-elevation forested areas. As well, Red-tailed Chipmunks are more arboreal than others of their kind, so finding them may involve scanning treetops.

These chipmunks may not hibernate in winter. They often remain awake in their dens, feeding lightly on their grand stores of food. If they do become dormant, it is only in a light torpid state that is easily broken.

DESCRIPTION: Like its kin, this large chipmunk has three dark and two light stripes on the face, and five dark and four light stripes on the back. The inner three dark stripes on the back are black; the dark facial stripes and the outermost dark stripes on the back are brownish. The rump is grayish. In keeping with its name, its tail is rufous above and brilliant reddish below, bordered with black and pale pinkish orange.

HABITAT: This chipmunk occurs in coniferous mountain forests and on boulder-covered slopes below treeline.

FOOD: Although conifer seeds, nuts, some berries and insects form most of the diet, it is not uncommon for chipmunks to feed on eggs, fledgling birds, young mice or even carrion.

DEN: As with all chipmunks, the Red-tailed Chipmunk usually spends winter in a burrow. Mothers often bear young in tree nests or cavities—the Red-tailed Chipmunk makes spherical tree nests more often than many other chipmunks.

YOUNG: Breeding occurs in spring, and after a one-month gestation, a litter of usually four to six young is born in May or June. The young are born blind and hairless. They grow rapidly and are usually weaned in about six weeks.

SIMILAR SPECIES: The Least Chipmunk (p. 202) is smaller, and in both it and the Yellow-pine Chipmunk (p. 204), the underside of the tail is grayish yellow, not brick red.

RANGE: The small range of this chipmunk includes only southeastern British Columbia, the very southwestern corner of Alberta, northeastern Washington, northern Idaho and western Montana.

Uinta Chipmunk

Tamias umbrinus

On the whole, all chipmunks look and behave very similarly, but the Uinta Chipmunk is one that stands out. Unlike most chipmunks, which prefer ground dwellings, the Uinta Chipmunk often builds its nest in a tree. As well, this deviant chipmunk puts on a large layer of fat to support itself through its winter hibernation. This unique chipmunk is named after the Uinta Mountains in the northeast corner of Utah, the heart of this species' range.

Uinta Chipmunks must be constantly alert for predators; threats can come from the sky, from the ground or even from below. Forest-dwelling raptors take a toll on chipmunk populations, as do foxes, Bobcats and snakes, but perhaps the most effective predators of chipmunks are weasels. Because weasels are small and streamlined, these voracious carnivores can sneak up on chipmunks from rock crevices or chase them into their burrows. Despite the relentless summertime predation, some Uinta Chipmunks have been known to live five to seven years in the wild, which is an unusually long time for a small rodent.

Like other chipmunks, the Uinta Chipmunk is finicky about keeping clean. A chipmunk's daily bathing involves cleaning its coat with its paws and feet. Once the coat is done, the chipmunk licks its paws clean, too. When necessary, chipmunks take "delousing" baths by rolling in dusty sand. In the process, some of the ticks or mites in their fur get dislodged by the dust and are left behind.

A chipmunk's meticulous cleaning is never enough to completely prevent parasites, however, and chipmunks are parasitized by botfly larva, tapeworms, fleas and lice. Sick or heavily parasitized chipmunks may be distracted and thus fall first to predators, which could help to keep parasites in check. Mammals that establish a nest in a fixed burrow system tend to accumulate fleas and other parasites with greater frequency and in larger numbers than do mammals that have no regular den site.

DESCRIPTION: This dark brownish chipmunk's lateral stripes are often so faint they may be difficult to discern against the dark brown or grayish-brown sides and rump. The belly is creamy white, with dark underfur. The ears are black with pale edges, and there is a grayish patch behind each ear. The tail is tawny below, bordered with brownish black and edged with cinnamon or pinkish buff. Above, it has a

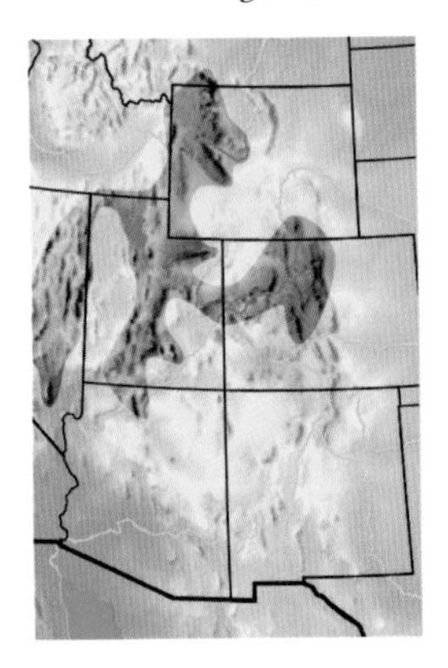

RANGE: The Uinta Chipmunk occurs in six or seven widely separated populations from the extreme southern edge of Montana south to Colorado, Utah and Nevada.

DID YOU KNOW?

This unusual, dark chipmunk seldom vocalizes, although females, like most mothers, may call to their recently emerged young.

Total Length: 7¾–9½ in (20–24 cm)

Tail Length: 3¾–4½ in (9.5–11 cm)

Weight: 2–3 oz (57–85 g)

background of brownish black mixed with tawny and edged with cinnamon or pinkish-buff hairs.

HABITAT: The Uinta Chipmunk occupies lodgepole pine and Douglas-fir habitats. It has a high elevation tolerance and can inhabit regions of subalpine fir. It has been seen in rocky outcroppings at the edge of the alpine tundra near the upper limits of tree growth.

FOOD: Conifer seeds, nuts and berries make up most of the diet. This chipmunk also eats a surprising amount of fungi and digs for buried mushrooms. Opportunistically, it will feed on insects or other animal matter.

DEN: The Uinta Chipmunk builds its nest in a burrow, often within the roots of a shrub, or uncommonly in a hollow log, tree cavity or roofed-over, abandoned bird nest. The nest chamber is about 7 in (18 cm) in diameter and is lined with dried grass and shredded bark.

YOUNG: Breeding takes place in spring. Only one litter of four to six young is produced a year, during late June or early July.

SIMILAR SPECIES: The Least Chipmunk (p. 202) is smaller. The Yellow-pine Chipmunk (p. 204) has less distinctive stripes on its sides. The Colorado Chipmunk (p. 206) has a generally grayer head, and its pale dorsal stripes are less distinct.

bounding group

Woodchuck

Marmota monax

For most of the year, Woodchucks are tucked quietly away more than 6½ ft (2 m) underground, relying on a lethargic metabolism to keep them alive during hibernation. They lie motionless, breathing an average of once every six minutes and maintaining life's requirements with a metabolic furnace fed by a trickle of fatty reserves. Once May returns (never as early as February's Groundhog Day), Woodchucks awake from their catatonic slumbers to breed and forage on the palatable new green shoots emerging with the warmer weather.

Woodchucks range across much of the northernmost Rockies, in low-mountain areas, finding shelter for their burrows in rock piles, under outbuildings and along riversides. In general, they are more solitary in nature than other kinds of marmots, and they are rarely seen far from their protective burrows, valuing security over the temptations of foraging. When Woodchucks do venture out to feed, it is often during the early twilight hours or shortly after dawn. Even then, they are wary and usually outrun most predators in an all-out sprint back the burrow. A shrill whistle of alarm typically accompanies a Woodchuck's disappearance into its burrow.

Historically, the Woodchuck lived in forested areas, and it can still be found in woodlands, although it now lives in great numbers on cultivated land—the Woodchuck is one of the few mammals to have prospered from human activity. Unhesitant about pilfering, Woodchucks living near humans often graze in sweet alfalfa crops to help fatten their waistlines. The luckiest Woodchucks find their way into people's backyards, where they stuff themselves on tasty apples, carrots, strawberries and other garden delights. In wild areas, Woodchucks follow the standard marmot diet of grasses, leaves, seeds and berries, sometimes supplemented with a bit of carrion.

ALSO CALLED: Groundhog.

DESCRIPTION: This short-legged, stout-bodied marmot is brownish, with an overall grizzled appearance. It has a prominent, slightly flattened bushy tail and small ears.

HABITAT: Woodchucks favor pastures, meadows and old fields close to wooded areas.

FOOD: This ground dweller eats primarily green vegetation and grasses and

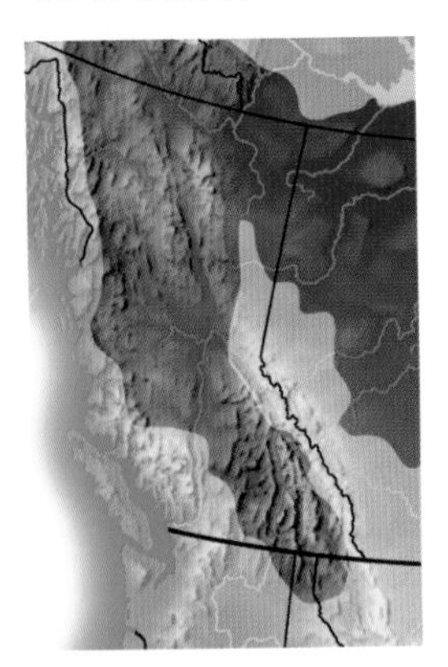

RANGE: The Woodchuck occurs from central Alaska east to Labrador and south to northern Idaho in the West and eastern Kansas, northern Alabama and Virginia in the East.

DID YOU KNOW?

Woodchucks are superb diggers that are responsible for turning over massive amounts of earth each year. As they burrow, they will periodically turn themselves around and bulldoze the loose dirt out of the tunnel with their stubby heads.

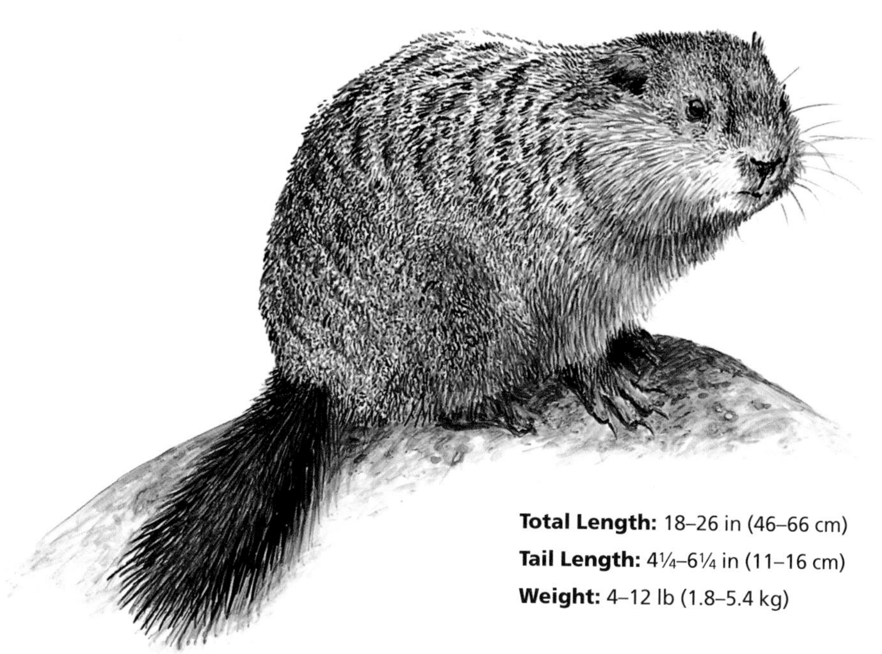

Total Length: 18–26 in (46–66 cm)

Tail Length: 4¼–6¼ in (11–16 cm)

Weight: 4–12 lb (1.8–5.4 kg)

some bark. The Woodchuck loves garden vegetables, and if it makes its way into urban areas, it may dine on corn, peas, apples, lettuce and melons.

DEN: The powerful digging claws are used to excavate burrows in areas of good drainage. The main burrow is 10–50 ft (3–15 m) long, and it ends in a comfortable, grass-lined nest chamber. A smaller chamber, separate from the nest chamber, is used for wastes. The dirt at the main burrow entrances is often populated by an assortment of plant species that are unique to these spoil piles. There are often plunge holes, without spoil piles, that lead directly to the nest chamber.

YOUNG: Mating occurs in spring, within a week after the female emerges from hibernation. After a gestation period of about one month, one to eight (usually three to five) young are born. The helpless newborns weigh only about 9 oz (260 g). In about four weeks their eyes open, and they look like proper Woodchucks after five weeks. The young are weaned at about 1½ months old. Their growth accelerates once they begin eating plants, and they continue growing throughout summer to put on enough fat for the winter hibernation and early spring activity.

SIMILAR SPECIES: The Hoary Marmot (p. 216) is generally gray and white with contrasting black markings, and it is typically found at higher elevations. The Yellow-bellied Marmot (p. 214) has, appropriately enough, a yellow belly, and in the Rockies it generally occurs farther south than the Woodchuck.

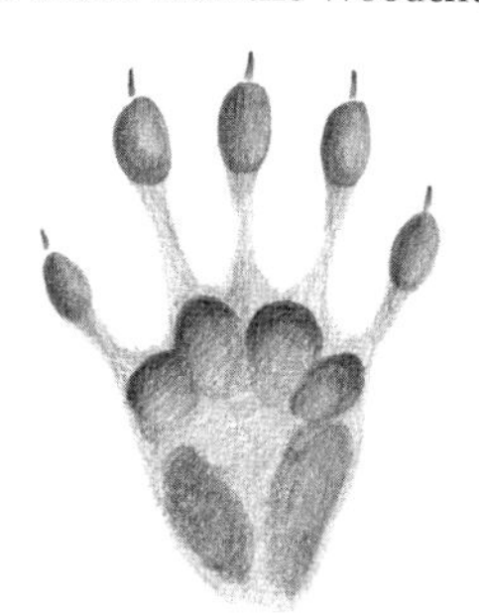

left hindprint

Yellow-bellied Marmot

Marmota flaviventris

True to its name, the Yellow-bellied Marmot has a distinct yellowish or burnt-orange belly. When this marmot is curious about or watchful of something, it often sits on its hindlegs in an upright position that displays its delightfully bright tummy.

Yellow-bellied Marmots sleep late, eat heartily and then snooze dreamily on warm rocks in the sun after they emerge from their burrows. Counting hibernation and nighttime sleep, Yellow-bellied Marmots spend about 80 percent of their lives in their burrows. They like their dens to be kept clean, and when they emerge from hibernation they throw out their used bedding and replace it with fresh grass and leaves. Throughout summer, they continue to keep their bedding clean and their burrows free of debris.

Colonies of Yellow-bellied Marmots have a strict social order, and whenever members of a colony are eating or wrestling with their family members, at least one marmot plays watchdog. This sentinel is responsible for warning the others if danger approaches. The alarm call is a loud chirp, which may vary in duration and intensity depending on the nature of the threat: short, steady notes probably translate as "Everyone pay attention, something's wrong"; loud, shrill notes convey the message "Into your burrows, now!" Different urgent warnings are reserved for immediate dangers, such as a circling eagle or an approaching fox.

Within the Rockies, marmot population sizes seem to be regulated by the availability of suitable hibernation sites. The dominant male of a colony evicts the younger males as they become sexually mature, and these banished marmots appear to suffer especially high overwinter mortalities.

ALSO CALLED: Rockchuck.

DESCRIPTION: The back is tawny or yellow-brown, grizzled by the light tips of the guard hairs. The feet and legs are blackish brown. The head has whitish-gray patches across the top of the nose, from below the ear to the shoulder and from the nose and chin toward the throat, which leaves a darker brown patch surrounding the ear, eye and upper cheek on each side of the face. The ears are short and rounded. The whiskers are dark and prominent. The dark, grizzled, bushy tail is often arched behind the animal and flagged from side to side. The bright buffy-yellow belly,

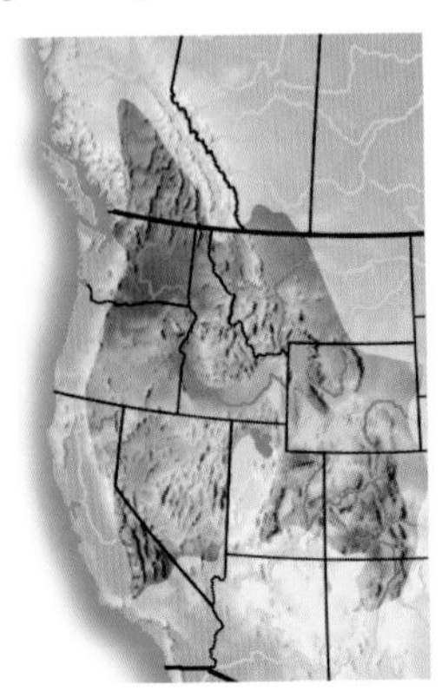

RANGE: Yellow-bellied Marmots are found from central British Columbia and extreme southern Alberta south into central California and northern New Mexico. They are especially common in Yellowstone National Park.

DID YOU KNOW?

Yellow-bellied Marmots frequently bask in the morning sun. At about midday they retire to their cool burrows, but in late afternoon they reemerge to feed. They seem to have poor control of their body temperature: in summer it may range from 93° F (34° C) to 104° F (40° C).

sides of the neck, upper jaw and hips are responsible for the common name.

HABITAT: Large rocks, either in the form of talus or outcrops, are a necessity, which accounts for this animal's alternate name "rockchuck." Within its range, this marmot may be found from valley bottoms to alpine tundra, but never in dense forests. It is a mountain animal south of Montana's Flint Creek Range; north of that it occurs in lowlands.

FOOD: Abundant herbaceous or grassy vegetation must be available within a short distance of the den. This marmot occasionally feeds on road-killed carrion, and there have been reports of the cannibalization of young.

DEN: Each of the adults maintains its own burrow, with those of the highest social status being nearest the colony center. A burrow is typically 8–14 in (20–36 cm) in diameter. It slants down for 20–39 in (51–99 cm) and then extends another 10–15 ft (3–4.6 m) to end beneath or among large rocks in a bulky nest lined with grass.

YOUNG: There are three to eight young in a litter, born in June after a 30-day gestation. Naked and blind at birth, they first emerge from the burrow at three to four weeks old. Well-fed females become sexually mature before their first birthdays. Males and females born at higher elevations usually do not get a chance to breed until they are at least two years old.

SIMILAR SPECIES: The Hoary Marmot (p. 216) has gray cheeks and a gray belly, and where their ranges overlap, it occupies higher elevations and rougher terrain. The Woodchuck (p. 212) is uniformly brownish, without the yellowish belly, and occurs farther north.

Total Length: 19–26 in (48–66 cm)

Tail Length: 5–7½ in (13–19 cm)

Weight: 3½–11 lb (1.6–5 kg)

Hoary Marmot

Marmota caligata

These stocky sentinels of alpine vistas pose graciously on boulders, gazing for untold hours at the surrounding Rocky Mountain scenery. They customarily emerge from their burrows soon after sunrise, but they remain hidden on windy days and during snow, rain or hailstorms.

Hoary Marmots exclusively occupy high-elevation environs, where long summer days allow rapid plant growth during a growing season that often lasts only 60 days a year. Despite the shortened summer season, these marmots seldom seem hurried; rather, most of their time seems to be spent staring off into the distance, perhaps on the lookout for predatory Grizzlies and Golden Eagles, or perhaps in simple appreciation of the spell-binding landscape.

Where they are frequently exposed to humans, Hoary Marmots can become surprisingly tolerant of our activities. These photo-friendly individuals are in sharp contrast to the wary animals that live in more isolated areas. In the backcountry, the presence of an intruder in an alpine cirque or talus slope is greeted by a shrill and resounding whistle, from which the nickname "whistler," *siffleur* in French, is derived. When alarmed, marmots travel surprisingly gracefully through boulder fields, quickly finding any one of their many escape tunnels.

Being chunky is most fashionable in Hoary Marmot circles. Although at first glance their surroundings may appear to hold few food possibilities, these high alpine areas are in fact rich in marmot foods. Marmots consume great quantities of green vegetation throughout summer, laying on thick layers of fat to be used during their eight- to nine-month hibernations. A considerable portion of stored fat remains when the marmots emerge from hibernation, but they need it for mating and other activities until the green vegetation reappears.

ALSO CALLED: Whistler.

DESCRIPTION: The head is gray and white with contrasting black markings. The cheeks are gray. A black band across the bridge of the nose separates the white nose patches from the white patches below the eyes. The ears are short and black. The underparts and feet are gray. A black stripe extends from behind each ear toward the shoulder. The shoulders and upper back are grizzled gray, changing to buffy brown on the lower back and rump, where black-tipped guard hairs surmount the under-

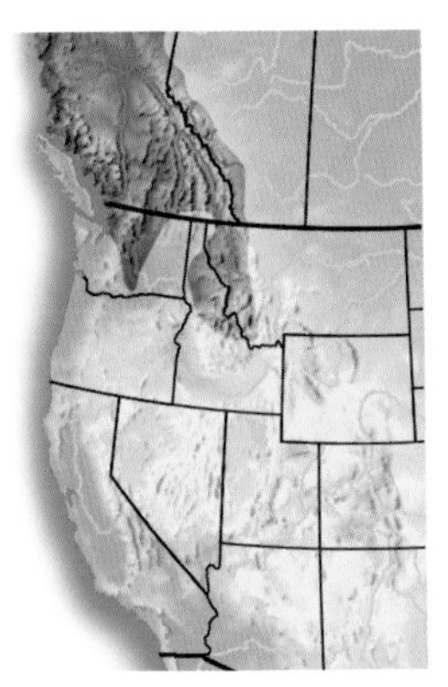

RANGE: The Hoary Marmot occurs from northern Alaska south through the mountains to southern Montana and central Idaho.

DID YOU KNOW?

Hoary Marmots often use nose-to-cheek "kisses" when greeting other colony members. Late morning, following avid feeding, is a peak period of socializing with other members. The kisses are shared between members of both sexes.

Total Length: 27–32 in (69–81 cm)

Tail Length: 7–9½ in (18–24 cm)

Weight: 11–15 lb (5–6.8 kg)

fur. The bushy brown tail is so dark it often appears black. This marmot often fails to groom its lower back, tail and hindquarters, so the fur there appears matted and rumpled.

HABITAT: Hoary Marmots are dependent upon large talus boulders or fractured rock outcrops near abundant vegetation in moist surroundings. They most commonly occur in alpine tundra and high subalpine areas.

FOOD: Such large quantities of many tundra plants are consumed that the vegetation near the burrows is often lawn-like from grazing. Grasses, sedges and broadleaf herbs are all eaten.

DEN: Burrows run about 6½ ft (2 m) into the slopes, where they may end as a cave, up to 3½ ft (1.1 m) in diameter, beneath a large rock. The nest chamber is often filled with soft grass.

YOUNG: A litter of four or five young is born in mid- to late May, about 30 days after mating. The fully furred young first emerge from the burrows in about the third week of July, when they weigh 7–11 oz (200–310 g). They are weaned soon after emerging and grow rapidly until they enter hibernation in September. Sexual maturity is achieved during their third spring.

SIMILAR SPECIES: The Yellow-bellied Marmot (p. 214) has a bright buffy-yellow belly and a grizzled brown back, and where the ranges of these two marmots overlap, the Yellow-bellied Marmot occurs at lower elevations.

Columbian Ground Squirrel

Spermophilus columbianus

From montane valleys to alpine meadows, the Columbian Ground Squirrel is a common resident of the Rocky Mountain region. Within its range, it seems that virtually every meadow has a population of this large rodent thriving among the grasses. At heavily visited day-use areas and campgrounds, colonies of this ground squirrel attract a surprising amount of tourist curiosity.

Columbian Ground Squirrels are robust, sleek and colorful animals that chirp loudly, often at the first sight of anything unusual. The chirps coincide with a flick of the tail and, in extreme cases, a split-second plunge down a burrow. The refuge-seeking behavior, however, is often preceded by a trill rather than a chirp. There are also different alarms for avian versus terrestrial predators, and for squirrel intruders from outside the colony. Making sense of the repertoire of different ground squirrel sounds may require more effort than most people are prepared to expend.

Colony members interact freely and non-aggressively with one another in most instances, sniffing and kissing their neighbors upon each greeting. The dominant male, however, which has his burrow near the center of the colony, maintains his central location through the breeding season. Ground squirrels from outside the colony are typically attacked by one or several members of the colony, and are driven far afield.

Dispersing individuals, forced to emigrate from their home colony to live, are exceedingly vulnerable to predation. Away from the sanctuary of communal life, these large rodents are a much-valued dietary choice for other mammals and birds. The Prairie Falcon, for one, may leave its grassland breeding grounds in late summer to concentrate on hunting Columbian Ground Squirrels in the mountains.

DESCRIPTION: The entire back is cinnamon buff, but because the dorsal guard hairs have black tips, a dappled, black-and-buffy effect results. The top of the head and the nape and sides of the neck are rich gray, with black overtones. There is a buffy eye ring. The nose and face are a rich tawny, sometimes fading to ochre-buff on the forefeet, but more frequently continuing tawny over the forefeet, underparts and hindfeet. The base of the tail is sometimes tawny or more rarely rufous. The moderately bushy tail is brown, overlain with hairs

RANGE: The Columbian Ground Squirrel is found from east-central British Columbia south to northeastern Oregon, central Idaho and western Montana.

DID YOU KNOW?

Columbian Ground Squirrels have been known to hibernate for up to 220 days. During hibernation, the squirrels wake at least once every 19 days to urinate and sometimes defecate, and to eat some stored food.

Total Length: 13–16 in (33–41 cm)

Tail Length: 3¼–4¾ in (8.3–12 cm)

Weight: 16–20 oz (450–570 g)

having black subterminal bands and buffy-white tips.

HABITAT: This wide-ranging ground squirrel may occupy intermontane valleys, forest edges, open woodlands, alpine tundra and even open prairies. Although it is primarily an animal of meadows and grassy areas, some individuals may learn to climb trees.

FOOD: All parts of both broadleaf and grassy plants are consumed. Carrion is eaten when it is found, and there are several reports of adults, especially males, being cannibalistic on young ground squirrels. Insects and other invertebrates are also eaten. Individual squirrels ordinarily store only seeds or bulbs in their burrows.

DEN: The colony develops its burrow system on well-drained soils, preferably loams, on north- or east-facing slopes in the mountains. The tunnels are 3–4¼ in (7.6–11 cm) in diameter and descend 3½–6½ ft (1.1–2 m). Each colony member's burrow system has 2 to 35 entrances and may spread to a diameter of more than 27 yd (25 m). A central chamber, up to 30 in (76 cm) in diameter, is filled with insulating vegetation. Several other burrows, each up to 5 ft (1.5 m) long, with a single entrance, serve as temporary refuges around the colony.

YOUNG: Mating occurs in the female's burrow soon after she emerges from hibernation, and after a gestation of 23 to 24 days, she delivers a litter of two to seven young. The upper incisors erupt by day 19, the eyes open at about day 20, and the young are weaned at about one month. All Columbian Ground Squirrels are sexually mature after two hibernation periods, but some females may mate after their first winter.

SIMILAR SPECIES: The Yellow-bellied Marmot (p. 214) is generally larger and has dark facial markings. The Richardson's Ground Squirrel (p. 220) does not have a tawny nose patch.

Richardson's Ground Squirrel

Spermophilus richardsonii

Maligned and vilified over much of its range, the Richardson's Ground Squirrel continues to be a common sight along the eastern edge of much of the Rockies, although its populations seem to be in a longterm decline. This ground squirrel often has a lengthy hibernation from August to February or March, but it is conspicuous, vocal and active during the day in the warmer months of the year. To many people, who often call it a gopher, this species is the most common ground squirrel, but in terms of overall North American distribution it has a smaller range than some other species. An overused name, "gopher" has been applied to several other burrowing rodents, as well as a tortoise and a snake.

Richardson's Ground Squirrels have earned quite a reputation for plundering grain crops and digging up farmland, an impressive accomplishment for creatures that spend 90 percent of their lives—counting nights, hot afternoons and eight months of hibernation—in a burrow. Few mammals in North America live such retired lifestyles.

Beyond their activities that conflict with agricultural practices, surprisingly little is known about the behavior of these prairie diggers. Only recently have studies unraveled the complexity of their hibernation and activity cycles. While all Richardson's Ground Squirrels are known to hibernate through much of the coldest winter weather, the males begin to emerge in spring, often before the last snows have retreated. They engage in prolonged courtship battles, aiming their bites at an opponent's testes. Many males do not survive the breeding season, and they seldom live more than two years. Courtship and reproductive duties are condensed between late spring and early summer, so that many adult ground squirrels have already begun to enter hibernation by the end of July. The first hibernators are typically the mature males, followed by the adult females, and then the juvenile females in late August. Any Richardson's Ground Squirrels observed later than that are almost certainly fat young males, which may remain active until October.

ALSO CALLED: Gopher, Picket Pin.

DESCRIPTION: The upperparts are a pinkish-buffy gray to cinnamon buff and are indistinctly mottled. The underparts are pale yellowish, pinkish or gray. The buffy-brown tail is about one-third

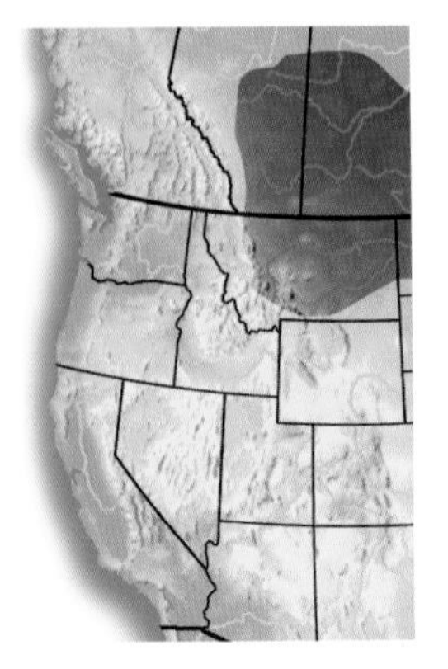

RANGE: This ground squirrel is mainly associated with the prairies and plains from southern Alberta east to southwestern Manitoba and south through Montana and to extreme western Minnesota.

DID YOU KNOW?

Richardson's Ground Squirrel burrows, which have an average of eight entrances, are used by dozens of other prairie mammals, amphibians, reptiles and invertebrates as hibernacula and refuge sites.

the length of the body, it is buffy to cream colored below, and it is fringed with short, black, white-tipped hairs. One of this ground squirrel's distinguishing characteristics is its habit of standing erect on its hindlegs to survey its surroundings, a trait that has earned it the name "picket pin."

HABITAT: An open country specialist, the Richardson's Ground Squirrel is common in prairies, meadows and pastures. It tends to avoid damp or fine-textured soils. It enters some montane and intermontane valleys along the eastern slopes of the Rockies.

FOOD: Like others of its kind, this ground squirrel eats flowers, fruits, seeds, grasses, green vegetation, insects and even some animal protein, mostly as carrion. When foraging, it stuffs its cheek pouches with seeds, which it carries back to its burrow for storage.

DEN: This ground squirrel often lives in loose colonies, particularly in favorable habitats. Families live in intricate burrows with the entrances marked at the side by large mounds of excavated dirt. The main burrow is about 15–30 ft (4.6–9.1 m) in length and ends in a grass-lined nest chamber. There are usually many secondary entrances and plunge holes.

YOUNG: Mating occurs after the female emerges from hibernation in spring. The gestation period is about 23 days, and the litter of about 3 to 11 (usually 7 or 8) young is born in May. The newborns are helpless, but they grow quickly and appear aboveground after about three weeks. They are weaned when they are about one month old, and their growth continues throughout summer as they prepare for their first hibernation.

SIMILAR SPECIES: The Columbian Ground Squirrel (p. 218) has a rich tawny nose. The Thirteen-lined Ground Squirrel (p. 226) has 13 alternately broken and solid, buffy lines running down its back.

Total Length: 11–13 in (28–33 cm)

Tail Length: 2⅜–3¼ in (6–8.3 cm)

Weight: 9⅛–22 oz (260–620 g)

Wyoming Ground Squirrel

Spermophilus elegans

The Wyoming Ground Squirrel was long considered to be a subspecies of the Richardson's Ground Squirrel, which it closely resembles. A detailed study of its biology, however, revealed that although these two species interbreed, their hybrids have reduced viability and do not survive long. In the field, the best way to distinguish between them is by differences in their vocalizations and in the color of the underside of the tail.

Although it is named after the "wild" state of Wyoming, this ground squirrel will never achieve the same iconic western status as bison, cowboys or lawlessness—its drab colors, high road fatalities and tendency to annoy ranchers diminishes its heroic appeal. For those with a discerning eye, however, these curious, tail-flicking squirrels are widespread through the Rocky Mountain states. Without them, the ecological and aesthetic make-up of this region would be significantly changed.

Wyoming Ground Squirrels live in colonies in areas that have fairly loose soils. The males tend to a have burrows near the colony center during the breeding season, but once the females bear young, the males move to the periphery or strike out on their own. Mothers share their burrows with their young, which eventually move to the edge of the colony by summer's end. These timely movements ensure that the colony is always experiencing inward and outward movements, so that the gene pool does not become saturated with local squirrels.

DESCRIPTION: The back is a grayish-buffy brown, the top of the nose is pinkish or cinnamon, and there is a light eye ring. The rump may have indistinct brown barring. In early summer, the sides and belly are yellowish buff. Later in the season they become grayer. The tail is grayish buffy to speckled black above; below it is pale orange or buffy. There is a short fringe of black outer hairs bearing buffy or whitish tips. The call is a weak, cricket-like alarm: *chirr*.

HABITAT: Dry slopes with sandy, gravelly or silty soils and short herbaceous vegetation are preferred. Wyoming Ground Squirrels never occur in dense forests, and in the Rockies they tend to be most frequently found in the foothills and mountain valleys.

FOOD: Green vegetation, especially sage and legumes, are important early

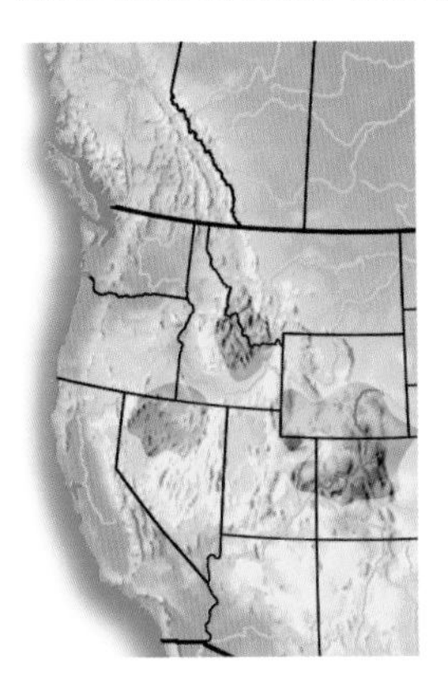

RANGE: The Wyoming Ground Squirrel is found in three separate areas: in the Rocky Mountains of northeastern Idaho and southwestern Montana; in northeastern Utah, southern Wyoming and central Colorado; and in southeastern Oregon and northern Nevada.

DID YOU KNOW?

The eyes of the Wyoming Ground Squirrel protrude so far to the side that this animal can see above its head, which is of value to an animal that is frequently the prey of hawks and eagles.

Total Length: 10–12 in (25–30 cm)

Tail Length: 2⅜–3⅛ in (6–7.9 cm)

Weight: 10–14 oz (280–400 g)

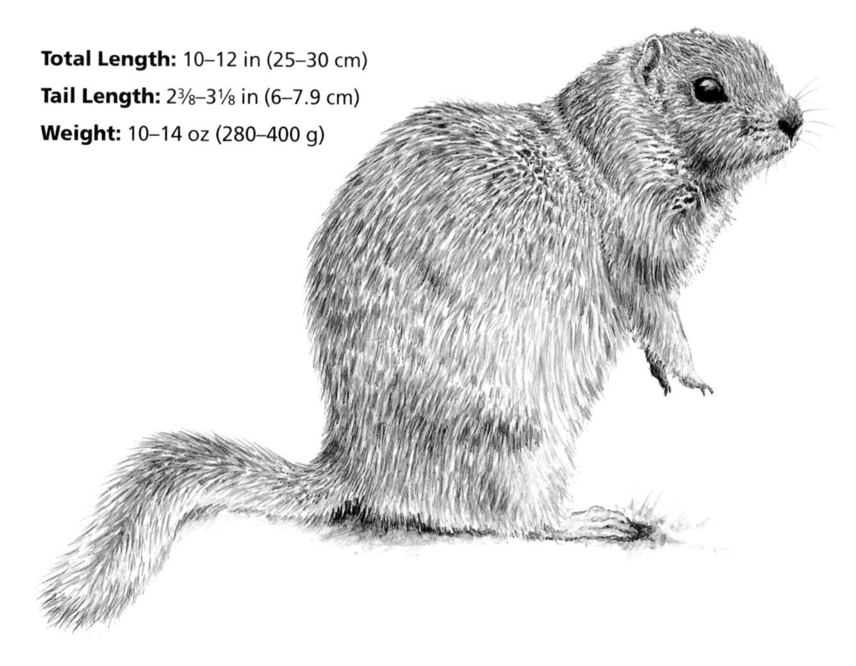

in the season. Some forbs and grass-like plants are ignored until they begin blooming or set seed, whereupon they become important in the diet. Insects, other invertebrates and carrion are often eaten.

DEN: The main burrow entrances typically have a mound of earth off to one side. Burrows may have up to eight entrances. Entrances that are dug from the inside have no external earth mound. The burrows are usually 4–11 yd (3.6–10 m) long and 2¾–3½ in (7–8.9 cm) in diameter. The burrow ends in a spherical nest chamber that is below frost line, about 9 in (23 cm) in diameter and filled with grass. Some food may be stored in the burrow.

YOUNG: Breeding begins a short time after the squirrels emerge from hibernation. Both yearling and adult females breed. A single litter of usually six to seven is born after a 22- to 23-day gestation. The eyes open and the upper incisors erupt when the young are about 23 days old. They then emerge from the burrow and switch to a diet of vegetation. Their growth accelerates, and they may gain up to ¼ oz (7 g) a day. They must gain about 10 oz (280 g) before they are ready for hibernation.

SIMILAR SPECIES: The Uinta Ground Squirrel (p. 224) is darker and larger, and the underside of its tail is gray instead of ochre-buff. The tail of the Richardson's Ground Squirrel (p. 220) is clay colored to buffy below.

running group

Uinta Ground Squirrel

Spermophilus armatus

Uinta Ground Squirrels know little of the surface appearance of their mountain homes outside their brief period of summer activity. Although these squirrels live within the core of the Rockies, they never see the fall colors, the great winter snows or the ptarmigans in their white winter plumage. Rather, their knowledge of their habitat is limited to the time between snow thaw and the end of summer. Yearling males have the longest active season, all of 97 days, while juveniles born in spring may spend only 55 days foraging aboveground before entering hibernation. In spite of this limited period of activity, Uinta Ground Squirrels are probably the ground squirrels most commonly seen by visitors to Yellowstone National Park.

The Uinta Ground Squirrel appears to fill the niche of its northern relative, the Columbian Ground Squirrel, in the southern and central Rocky Mountains. Perhaps mountain populations of both of these ground squirrels share an inability to withstand high temperatures. The name "Uinta" is derived from the location of the first collected specimen: the Uinta Mountains near Fort Bridger, Wyoming.

Although these attractive, brightly colored squirrels can sit upright, they do not do so as often as Richardson's Ground Squirrels. Instead, Uinta Ground Squirrels may lie motionless, feeding in one area for some time before rushing to a new area to begin feeding again. Often, they tend to concentrate their feeding activity in the vicinity of large rocks, which may afford them some protection from overhead raptors. Where food is abundant, Uinta Ground Squirrels live in large, dense colonies. They excavate complex burrow systems that house the colony members. Considerable amounts of earth are overturned by these diggers, which thereby contribute to soil aeration and nutrient cycling. Burrows are often constructed in areas of winter snow accumulation on north-facing slopes.

DESCRIPTION: The head, front of the face and ears are cinnamon, with gray highlights on the crown. The sides of the neck and face are pale gray. There is a light buffy eye ring. The back is cinnamon buff, and the dorsal hairs have light pinkish-buff tips. The sides are paler and more buffy than the back. The feet and belly are even lighter pinkish buff to buffy white. The underside of the tail is gray, and intermingled black and buffy-white hairs on the tail make it quite dark.

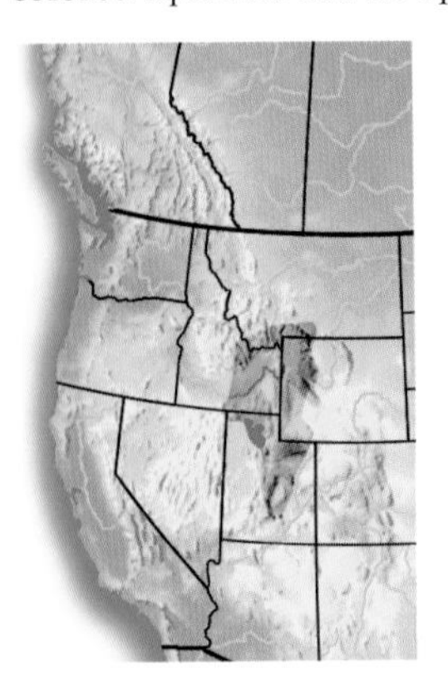

RANGE: The Uinta Ground Squirrel has a range that extends south from southern Montana through western Wyoming, eastern Idaho and north and central Utah.

DID YOU KNOW?

Adult male Uinta Ground Squirrels enter hibernation as early as mid-July. This period of dormancy continues until they emerge between late March and mid-May, ready to begin a new period of activity.

Total Length: 11–12 in (28–30 cm)

Tail Length: 2⅜–3 in (6–7.6 cm)

Weight: 10–15 oz (280–430 g)

HABITAT: The Uinta Ground Squirrel prefers well-drained subalpine and streamside meadows and edge areas between meadows and forests. It generally avoids dry, shortgrass prairies. The areas it occupies are nearly always at elevations above 5200 ft (1580 m), so this squirrel is truly an inhabitant of the mountains and high plateaus.

FOOD: This squirrel appears to be nonselective, feeding on both broad-leaved and grass-like plants. Some insects and carrion are eaten.

DEN: The complex burrows of Uinta Ground Squirrel colonies may have up to eight entrances. Some entrances have a mound of earth to one side, but others are dug from inside and have no external mound. Burrows can be up to 11 yd (10 m) long, ending in a nest chamber lined with grass. There are no records of food being stored within the burrow.

YOUNG: Yearling and older females come into estrus within a few days of emerging from hibernation, and mating takes place in the female's burrow. Usually 4 to 6 young, weighing an average 1/4 oz (7 g), are born in the grassy burrow nest. The eyes open after 18 days, and the young emerge from the burrow at about three weeks old.

SIMILAR SPECIES: The Wyoming Ground Squirrel (p. 222) is lighter overall and has ochre-buff, rather than gray, on the underside of its tail.

Thirteen-lined Ground Squirrel

Spermophilus tridecemlineatus

Although the Thirteen-lined Ground Squirrel occurs throughout much of the Rocky Mountains, it is an animal that can easily be missed. This small, striped squirrel seems easy to distinguish from the far more common Richardson's Ground Squirrel, but its spots and stripes serve to camouflage this animal, making it more difficult to see. Be sure to thoroughly scan all lone, slender roadside squirrels so that you don't overlook any of the 13-lined variety. Where one Thirteen-lined Ground Squirrel is found, others are likely present, although they rarely exhibit the same degree of colonialism as Richardson's Ground Squirrels.

Thirteen-lined Ground Squirrels tend to favor areas with taller grass, through which they cut paths reminiscent of voles' runways. The squirrels' striped backs blend perfectly with the alternating pattern of sun and shade created by the tall grass. They usually move in a series of rushes interspersed with stops of irregular length. When alarmed, these rodents utter a shrill *seek-seek* or a high-pitched trill. They often stop short of entering their burrows, posing near the entrance instead and often allowing an observer to approach closely before disappearing from sight.

The Thirteen-lined Ground Squirrel is one of the most predacious ground squirrels in the Rocky Mountains. Its diet is primarily vegetarian upon emergence from hibernation, but it shifts markedly during late May and June, when animal matter can make up almost half the daily intake. The Thirteen-lined Ground Squirrel may even climb up into small trees or shrubs in search of promising bird nests. By the end of summer, these striped beasts have consumed so much food that they are fat, irritable and ready for bed. The listless squirrels curl into tight balls in their solitary nest chambers, where they will stay for about seven months.

DESCRIPTION: The brownish back bears 13 alternately dotted and solid, buffy stripes. The head appears long and narrow, with large eyes and small ears. The top of the head is buff, sprinkled with brown. The eye ring, nose, cheeks, feet and underparts are buffy. The sides are gray. The tail is cylindrical, not bushy, and its central color is tawny. The longest tail hairs have a blackish subterminal band and a buffy tip.

HABITAT: This squirrel favors brushy edges of tall grass with nearby herba-

RANGE: Occurring through much of central North America, this ground squirrel ranges from east-central Alberta east through Minnesota to Ohio and south to northern New Mexico, southeastern Texas and Missouri.

DID YOU KNOW?

True to its name, there actually are 13 buffy lines on this animal's back, although seven of the "lines" are rows of light spots.

Total Length: 8¼–12 in (21–30 cm)

Tail Length: 2⅝–4¼ in (6.7–11 cm)

Weight: 2⅞–8⅛ oz (82–230 g)

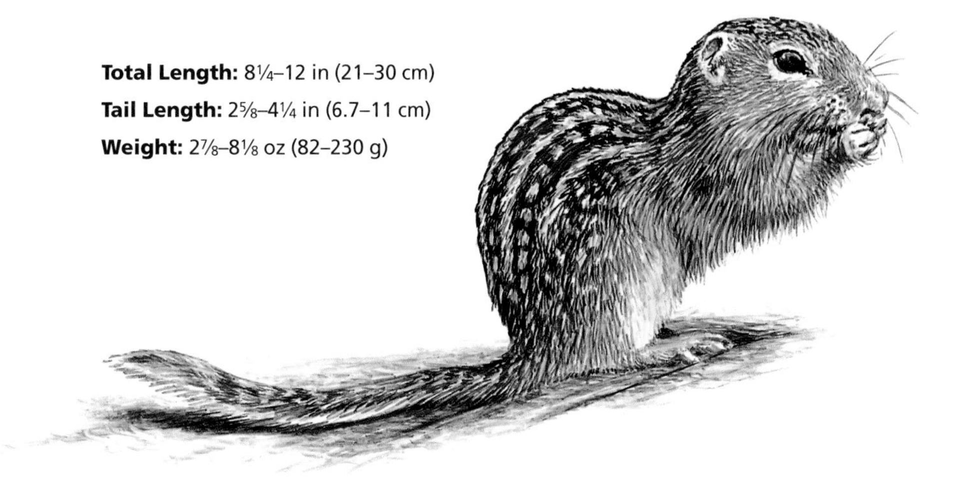

ceous vegetation. Although it is most common at lower elevations, it ranges into the eastern slopes of the Rockies in southern Alberta and Montana, and it is found in some parts of the mountains of Wyoming and Colorado.

FOOD: Seeds seem to be the staple component of the diet, but the Thirteen-lined Ground Squirrel eats more meat items than other ground squirrels. Insects and other invertebrates, young birds, mice and carrion are sought out and devoured. This squirrel also eats berries and native fruits, and it is sometimes a garden pest, consuming peas, beans, strawberries and melons.

DEN: There is almost never an earth mound near the burrow entrance. The tunnel, which is 2–2½ in (5.1–6.4 cm) in diameter, descends steeply for 4–40 in (10–100 cm) and then typically makes a right-angle bend and levels out. The maximum reported diameter of a burrow system is 20–30 ft (6.1–9.1 m). The nest chamber, about 9 in (23 cm) in diameter, is up to 6 ft (2 m) down in a passageway off the main burrow. It is typically filled with fine grass and dried roots.

YOUNG: Thirteen-lined Ground Squirrels mate soon after the females emerge from hibernation. After a gestation period of 27 to 28 days, a litter of usually 8 to 10 naked, blind and helpless young is born. Soon after their eyes open at 26 to 28 days, the young emerge from the burrow and switch to a diet of vegetation and meat. They are sexually mature following their first hibernation.

SIMILAR SPECIES: The Richardson's Ground Squirrel (p. 220) and the Wyoming Ground Squirrel (p. 222) are larger and neither has a striped back.

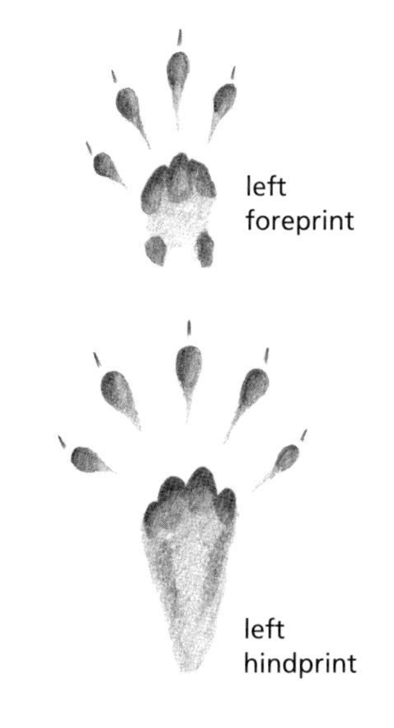

Idaho Ground Squirrel

Spermophilus brunneus

Total Length: 8¼–8⅝ in (21–22 cm)

Tail Length: 1¾–2 in (4.5–5.1 cm)

Weight: 3½–4⅝ oz (99–131 g)

Sporting a white chin and golden highlights on its upperparts, the Idaho Ground Squirrel is the rarest ground squirrel in North America. It is even harder to find this rodent because it is awake and active for only one-third of the year. An Idaho Ground Squirrel enters its burrow in late July or August, when the weather is often too hot and dry for its liking. Its dormancy continues until the following spring, nearly eight months later. When this squirrel finally emerges from hibernation, it immediately begins looking for food and a mate.

DESCRIPTION: This small ground squirrel is mainly dappled brown or gray over its upperparts. It has distinct golden highlights and a reddish nose. Its chin is white, and its tail has five to eight alternating dark and light bands around the edge.

RANGE: The Idaho Ground Squirrel is found only in a small part of west-central Idaho.

HABITAT: This rare squirrel may occupy montane valleys and open grasslands and meadows surrounded by forests of ponderosa pine.

FOOD: All parts of both broadleaf and grassy plants are consumed, especially seeds and onion bulbs. Carrion is eaten when it is found, and insects and other invertebrates are also eaten.

DEN: The colony develops its burrow system on well-drained soils. The burrows can become very complex, with each colony member's burrow having 2 to 35 entrances. A central chamber is filled with insulating vegetation. Several other burrows, each up to 5 ft (1.5 m) long and sporting just a single entrance, serve as temporary refuges around the colony.

YOUNG: Mating occurs in the female's burrow soon after she emerges from hibernation. After a gestation of 22 to 24 days, she delivers a litter of two to seven young. Very little is known about the development of the young.

SIMILAR SPECIES: The Columbian Ground Squirrel (p. 218) is much larger and redder. The Golden-mantled Ground Squirrel (p. 230) has distinct stripes down its sides.

Rock Squirrel

Spermophilus variegatus

Total Length: 17–21 in (43–53 cm)

Tail Length: 6⅜–10 in (16–25 cm)

Weight: 21–28 oz (600–790 g)

With its long, bushy tail, the Rock Squirrel looks a lot like a tree squirrel, and, more so than other ground squirrels, it is an agile climber that often seeks out berries and seeds in trees and shrubs. The Rock Squirrel also goes against the ground squirrel norm by only having a short, intermittent hibernation period. Much to a northern naturalist's surprise, it is not uncommon to see a Rock Squirrel abroad in December.

DESCRIPTION: The back is grayish to blackish gray and often appears mottled. The shoulder area is darker than the rump. The sides and upper surfaces of the feet are brownish. The belly is a light grayish brown. The bushy tail is rarely as long as the body.

HABITAT: True to its name, this squirrel inhabits rocky areas, primarily on low cliffs, canyon walls, boulder piles and talus slopes. It is sometimes encountered along brushy streamsides.

FOOD: Seeds and fruits seem to be the staple component of the diet, but Rock Squirrels also eat insects, other invertebrates, young birds, eggs and carrion.

DEN: Rock Squirrels form large colonies that have a distinct social order. Females make their burrows in the center of a colony's territory, while one dominant and a few subordinate males burrow on the outskirts. The burrows are dug underneath large boulders, which provide excellent protection against predators. The entrances are located in rock crevices, and the burrows may be up to 11 yd (10 m) long. A burrow ends in a spherical nest chamber that is filled with insulating grass. Some food may be stored in the burrow.

YOUNG: A litter contains three to nine young, which weigh about 1/4 oz (7 g) at birth. Unlike other ground squirrels, Rock Squirrels may have two litters a year in some places, one in May or June and another in August or September. In Colorado there is only a spring litter.

SIMILAR SPECIES: No other large, bushy-tailed ground squirrel is found in the same range. The Columbian Ground Squirrel (p. 218) is more colorful.

RANGE: The Rock Squirrel is found from northern Colorado west to southern Nevada and south into Mexico through Arizona and western Texas.

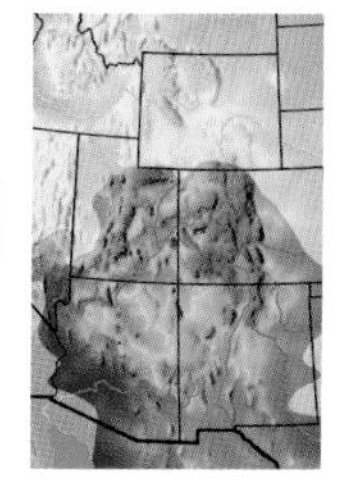

Golden-mantled Ground Squirrel

Spermophilus lateralis

In spite of this squirrel's often bold behavior and curiosity toward human visitors in the mountain parks, it is frequently the victim of mistaken identity. Misled by its long white and black side stripes, onlookers often call this small ground squirrel a chipmunk. Closer inspection, however, reveals the stripes' distinct character: they stop short at this ground squirrel's neck. All chipmunks have stripes running through their cheeks.

Golden-mantled Ground Squirrels are attractive, stocky, brown-eyed charmers. Bold, buffy-white eye rings frame their endearing eyes, which seem to give expression to their antics. These squirrels are found alongside pikas in talus slopes, where they both continually appear and disappear among the boulders. If you can imitate the high-pitched cries of either of these two animals, the ground squirrels may approach, suddenly appearing perched on a rock surprisingly close by. At close range, you can often see their bulging cheek pouches crammed with seeds and other foods, ready to be stored in their burrows.

Although Golden-mantled Ground Squirrels, which are often common around campsites or picnic areas, frequently mooch handouts from visitors, feeding them (or any other wildlife) is illegal in the national parks. Human handouts often lead to extreme obesity in animals, which is unhealthy. Perhaps when visitors to the parks become better informed, they will resist the temptation to feed these and other "friendly" animals, instead satisfying their nurturing instincts with detailed observations and quiet awe.

DESCRIPTION: The head and front of the shoulders are a rich chestnut. The buffy-white eye ring is broken toward the ear. Two black stripes on either side of a white stripe run along each side from the top of the shoulder to near the top of the hip. The back is grizzled gray. The belly and feet are pinkish buff to creamy white. The top of the tail is blackish, bordered with cinnamon-buff. The lower surface of the tail is cinnamon-buff in the center. There is a black subterminal band and a cinnamon-buff fringe on the hairs along the edge of the tail.

HABITAT: This squirrel inhabits montane and subalpine forests wherever rock outcrops or talus slopes provide adequate cover. At least in summer, if not always, low numbers reside in or beside the alpine tundra.

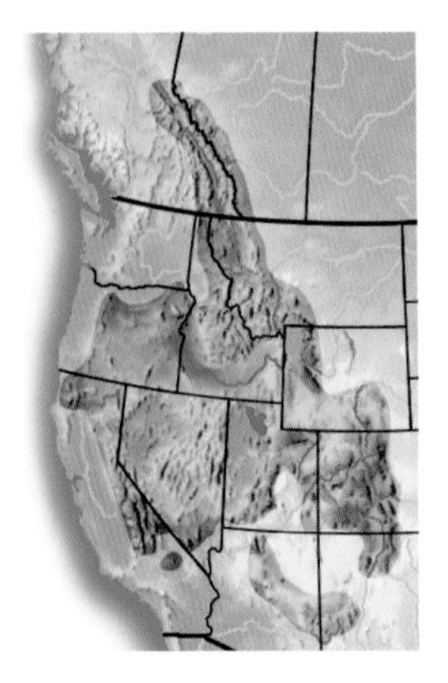

RANGE: This rock-dwelling squirrel's range is restricted to the Rocky Mountains and the southern Cascades and Sierra Nevada.

DID YOU KNOW?

A Golden-mantled Ground Squirrel can cram so much material into its cheek pouches that its head seems to have two huge, swollen tumors. This squirrel will work, practically without stopping, to carry away an entire supply of oats or peanuts to its secret stash.

GOLDEN-MANTLED GROUND SQUIRREL

Total Length: 11–13 in (28–33 cm)

Tail Length: 3¾–4¾ in (9.5–12 cm)

Weight: 6–12 oz (170–340 g)

FOOD: Green vegetation forms a large part of the early summer diet. Later, more seeds, fruits, insects and carrion are eaten; still later, conifer seeds become a major component of the fall diet. Fungi are another common food.

DEN: The burrow typically begins beneath a log or rock. The entrance is 3 in (7.6 cm) in diameter and lacks an earth mound. The tunnel soon constricts to 2 in (5.1 cm), and although most burrows are about 3½ ft (1.1 m) long, others may extend to 15 ft (4.6 m). Two or more entrances are common. The nest burrow ends in a chamber that is 6 in (15 cm) in diameter and has a mat of vegetation on the floor. Nearby blind tunnels serve as either latrine or food storage sites. Like many ground squirrels, this species closes the burrow with an earth plug upon entering hibernation and sometimes when it retires for the night.

YOUNG: Breeding follows soon after the female emerges from hibernation in spring. After a gestation period of 27 to 28 days, four to six naked, blind pups are born between mid-May and early July. At birth, the young weigh about ⅛ oz (3.5 g). The eyes open and the upper incisor teeth erupt at 27 to 31 days. The young are weaned when they are 40 days old. They enter hibernation between August and October and are sexually mature when they emerge in spring.

SIMILAR SPECIES: Chipmunks (pp. 202–10) are much smaller, and their stripes extend through the face.

White-tailed Prairie-Dog

Cynomys leucurus

Easily identified by their white-tipped tails, these prairie-dogs occur in great numbers in the sagebrush plains of Utah, Wyoming and Colorado. Because they live at high, cool elevations where winters can be cold and long, these ground dwellers must hibernate. They sleep straight through from late October to March. When they emerge after a long winter, they are slim and hungry, but after a few good meals of spring seedlings, these prairie-dogs are ready to mate.

White-tailed Prairie-Dogs are less sociable with each other than Black-tailed or Gunnison's prairie-dogs: they live in smaller colonies and spend less time engaged in grooming and "kissing" behaviors. Unlike their black-tailed cousins, they do not engage in cooperative excavation projects to link each burrow to its neighbors. For safety and social order, these prairie-dogs may excavate just a few links in the passageways of their town. The burrows also tend to be more dispersed, with younger animals living on the periphery of the colony and dominant ones at the core. Colony members warn of danger with a chatter-like call unlike that of other prairie-dogs.

White-tailed Prairie-Dog burrows are marked with large mounds of dirt that can be up to 3 ft (91 cm) high and 9 ft (2.7 m) across. Because of their extremely large size, however, the mounds may be overlooked as entranceways to the prairie-dog town. The burrows are generally located in stands of shrubs or brushy vegetation near abundant supplies of grasses and sedges.

From the time that they emerge from their burrows in spring until they enter winter dormancy in late fall, White-tailed Prairie-Dogs must at least double their body weight. Over the course of the summer, they fatten themselves on grasses and herbaceous plants; by October, their bellies wax large and they become steadily lazier. As their weight increases and their metabolism slows, they become quite irritable and less inclined to tolerate visitors of any sort.

DESCRIPTION: The back is yellowish buff, as is the nose. There are prominent blackish-brown patches above each eye and on each cheek. The ears are pale cinnamon. Black-tipped guard hairs on the back wear unevenly, creating a banded, mottled or speckled appearance in different individuals. The limbs, feet and belly are uniformly buffy. The claws are black with light tips. The end of the tail is all white; the basal half has hairs

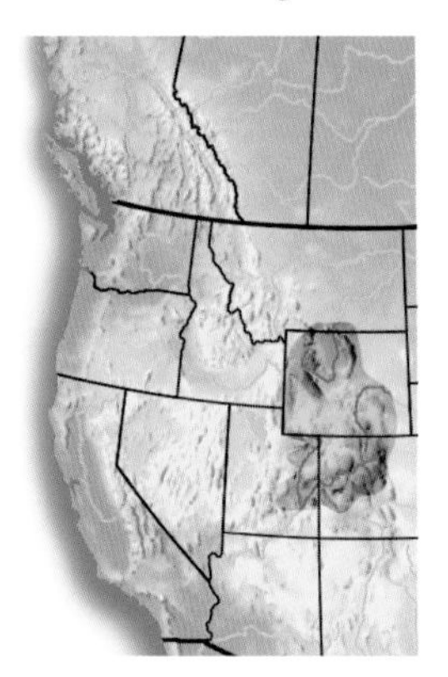

RANGE: This prairie-dog's range extends from extreme southern Montana through western Wyoming, west-central Utah and northwestern Colorado.

DID YOU KNOW?

White-tailed Prairie-Dogs are often the favorite prey of Golden Eagles. In a single season, 90 percent of the young, plus a few adults, were taken from one colony by these eagles.

Total Length: 13–15 in (33–38 cm)

Tail Length: 1¾–2⅜ in (4.5–6 cm)

Weight: 1½–2¾ lb (680–1250 g)

with mixed white and blackish tips over a cinnamon base. Sometimes the entire animal is stained the color of the soil where it lives.

HABITAT: The White-tailed Prairie-Dog typically occupies mountain meadows at higher elevations, but at the northern limit of its range, colonies may be found in semi-desert communities adjacent to rivers.

FOOD: Leaves, stems and a few roots make up most of the diet. Insects and carrion are sometimes consumed, but sedges and grasses are of paramount importance.

DEN: Burrow entrances are usually surrounded by a diffuse mound. The burrow descends at an angle of about 40°, instead of vertically. Burrows are 7 in (18 cm) in diameter, may be 10 ft (3 m) long and descend 7 ft (2.1 m) deep. Occasionally there are branch tunnels. A cavity measuring 18 x 11 in (46 x 28 cm) at the tunnel end serves as the hibernation site. Feces are used to plug the tunnels prior to hibernation.

YOUNG: Females mate soon after their emergence in March. Following a gestation period of about 30 days, usually five to six naked, blind and helpless pups are born. By mid-July the young are independent and seeking their own burrows. They are sexually mature the following spring.

SIMILAR SPECIES: The Gunnison's Prairie-Dog (p. 234) has a darker back and gray, not white, hairs on the tip of its tail, and its alarm call is a melodious, single-syllable call repeated many times.

Gunnison's Prairie-Dog

Cynomys gunnisoni

Their high-pitched barks and bubbly chatter give away the presence of Gunnison's Prairie-Dogs. Living on the high grassland slopes and plateaus of the southern Rocky Mountains, these prairie-dogs behave more like ground squirrels than other members of their genus.

While Black-tailed Prairie-Dogs can live in extremely large towns on the prairies, Gunnison's Prairie-Dogs generally live in colonies of about 50 individuals. The size of a colony is strongly affected by the surroundings. In flatter habitats, colonies can be much larger because each member of the group can see the others, which strengthens the safety of the colony and maintains hierarchies. In less desirable habitats, such as shrubby terrain or rolling hills, increased predation and the lack of suitable burrowing ground prevent large colonies from forming. Human encroachment in this prairie-dog's preferred territories has pushed it farther into marginal habitats.

Although their colonies tend to be larger, the burrows of Gunnison's Prairie-Dogs are less complex than those of White-tailed Prairie-Dogs. Each burrow has only one nest chamber, and it does not have any food or waste alcoves. Active only during the day, these prairie-dogs eat and defecate outside. Gunnison's Prairie-Dogs excavate their burrows on slightly sloped land to maximize water drainage. They do not tamp the excavated dirt into a mound at the entrance to the burrow, nor do they remove grass and weeds that sprout near the opening; instead, loose rocks, grass tufts and scattered dirt disguise the whereabouts of this prairie-dog's home.

DESCRIPTION: The back is yellowish buff, with intermixed blackish hairs. The belly is lighter and has fewer black hairs. The cheeks, sides of the head and eyebrows are often darker than the rest of the head. The end of the tail is grayish white.

HABITAT: Gunnison's Prairie-Dogs inhabit open, grassy and brushy areas at relatively high elevations.

FOOD: These prairie-dogs graze grasses, sedges, forbs and shrubs. They do not appear to dig or eat roots, and they often eat only the fruits and flowers of plants. They may graze the vegetation around their burrows to a very short, lawn-like appearance.

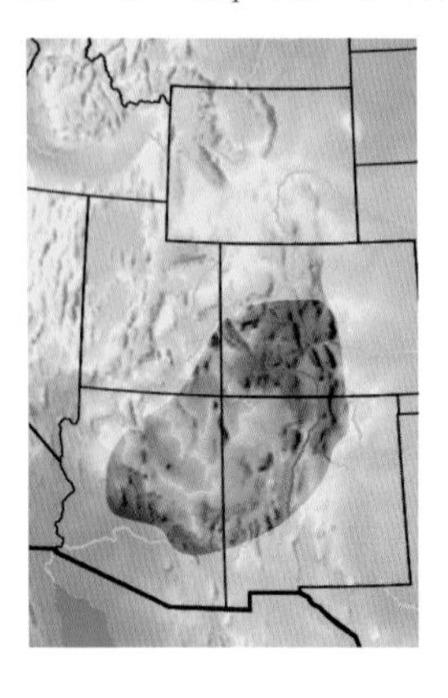

RANGE: This prairie-dog, which has a very limited distribution, occurs in Arizona, New Mexico, Colorado and Utah. Rodent eradication campaigns have eliminated it from much of its former range, and small, scattered colonies are the rule today.

DID YOU KNOW?

The danger call of the Gunnison's Prairie-Dog is a series of high-pitched barks. As the danger becomes more threatening, the call increases in intensity and speed.

Total Length: 12–15 in (30–38 cm)

Tail Length: 1½–2⅜ in (3.8–6 cm)

Weight: 23–42 oz (650–1190 g)

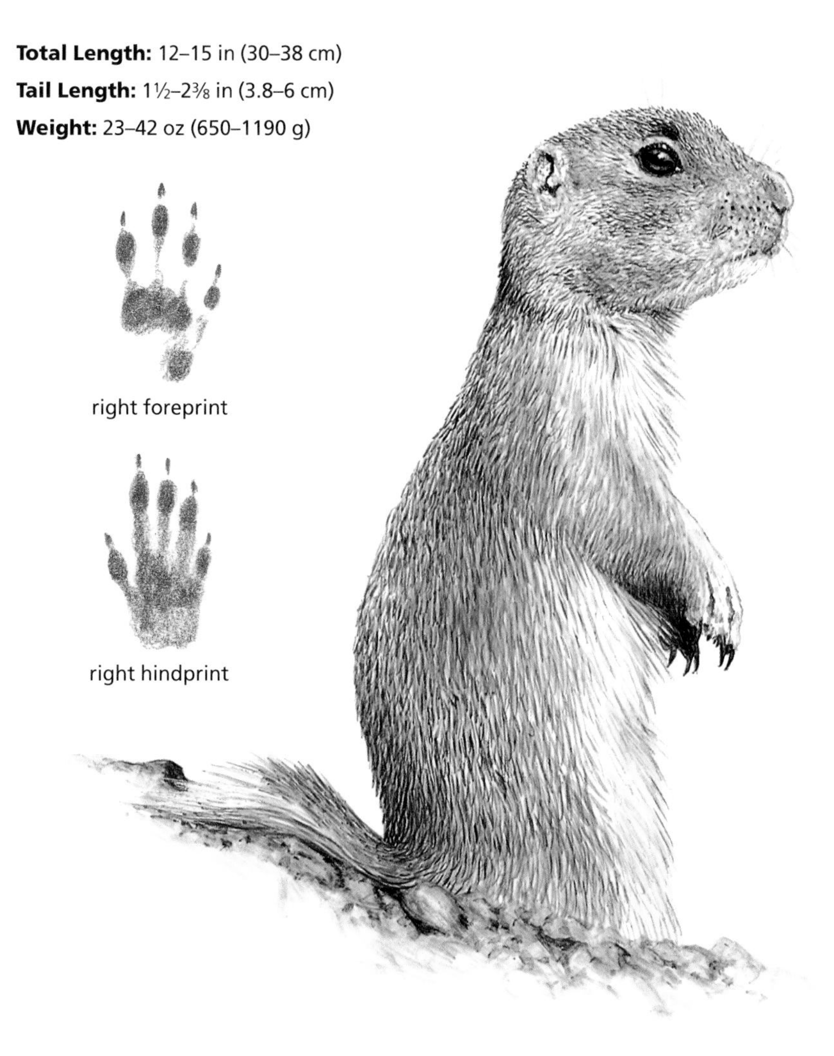

right foreprint

right hindprint

DEN: The burrows are often widely dispersed on only slightly sloping land. The mounds around the burrow entrances are seldom large, and the animals do not work the mounds and make little effort to keep them free of vegetation. Loose dirt and plants help hide the whereabouts of burrow entrances.

YOUNG: Mating occurs in April or May, soon after the females emerge from hibernation. The young are born slightly more than 30 days later, and, following the pattern of ground squirrels, they first emerge from the burrow four to five weeks later, usually in early July. There is one litter of three to eight pups each summer. Gunnison's Prairie-Dogs are sexually mature after their first hibernation.

SIMILAR SPECIES: The White-tailed Prairie-Dog (p. 232) is generally larger and has a white-tipped tail.

Eastern Fox Squirrel

Sciurus niger

The range of the Eastern Fox Squirrel historically lay well east of the Rocky Mountains, but through natural expansion and both deliberate and accidental introductions, it is expanding westward up to the eastern slopes of the Rockies, and it occurs in some mountain valleys.

The largest and most colorful of the tree squirrels, the Eastern Fox Squirrel comes in three different color forms: in the northeastern part of its range, it is dusty gray with yellowish undersides; moving west, its undersides become bright tawny red; in southern regions, it is often black, with a silvery rump and tail and a white blaze on the face. Only the gray (mostly) and black phases are seen in the Rocky Mountains.

Eastern Fox Squirrels usually lead solitary lives, but in areas where food is plentiful, many of these squirrels gather to feed. They collect seeds, fruits, fungi, green pinecones and corn. They bury non-perishables, such as seeds and nuts, in caches just under the ground surface. Their most gregarious behavior occurs in winter, when several adults whose home ranges overlap, and which are often related, share food caches and tree cavities. The sharing that occurs between Eastern Fox Squirrels is not as sociable as the behavior seen in ground squirrels: fox squirrels do not groom each other, nor do they "kiss" or nuzzle to maintain friendly ties. Eastern Fox Squirrels remain active through winter, and the individuals sharing a tree cavity come and go regardless of the cold.

Where tree cavities are scarce, fox squirrels build medium-sized tree nests. Just over 1 ft (30 cm) wide, the nests are spherical masses of leaves and twigs built in tree forks high off the ground. The nest materials are taken from the tree in which they are built. Green and leafy, the nests are hard to distinguish amid the foliage. Throughout their home range, mature squirrels usually maintain three to six nests.

DESCRIPTION: Most fox squirrels seen in the Rockies are of the gray phase. They have a grizzled-gray back with buffy undertones. The feet, belly and cheeks are buffy to cinnamon, and the insides of the ears are orangish. There is a buffy eye ring and a tawny patch behind each ear. The long, bushy tail has orange below, then black, then a buffy border. The top of the tail is nearly black, with buffy overtones. The claws and whiskers are black. Fox squirrels of the black color phase, which are

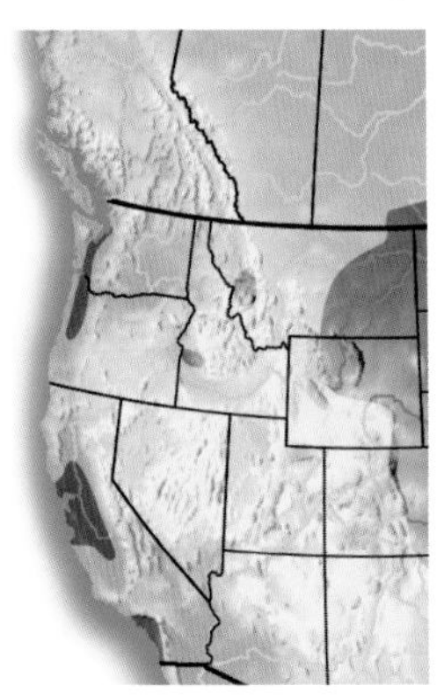

RANGE: The Eastern Fox Squirrel occurs through most of the eastern U.S., except New York and New England, typically as far west as eastern Montana and western Texas. Introduced populations inhabit several places in the West.

DID YOU KNOW?

While the food caches of other squirrels may germinate if left for too long, fox squirrels have been known to nip off the germinating ends of nuts before burying them. When they return to their caches weeks or months later, the ungerminated nuts are an easy meal.

Total Length: 20–23 in (51–58 cm)

Tail Length: 8–11 in (20–28 cm)

Weight: 1¼–2½ lb (570–1130 g)

uncommon in the Rockies, are almost entirely black.

HABITAT: Characteristically, fox squirrels inhabit open deciduous woodlands and forest edges with abundant mast-producing trees. Fox squirrels have been introduced into many cities and have dispersed into cottonwood forests along river valleys and residential areas where introduced trees have matured. They prefer a mixture of open areas and forests.

FOOD: Favorite foods are nuts and acorns, but these squirrels enjoy corn, the fruits of elm, ash, cherries, plums and hawthorn, plus grain and mushrooms. In spring they may chew the bark of maples and box elders to feed on the sweet cambium.

DEN: These squirrels usually make a home by modifying woodpecker nests in tree cavities. Leaf nests are constructed and sometimes are used throughout winter. Rarely, the nest is located in a rotten stump or root below the ground.

YOUNG: In the northern portion of the range there is a single litter each year, but a few squirrels further south bear a second litter in late summer. Most breeding is in December and January, followed by a 45-day gestation period. Usually three to four blind, helpless young are born. Their eyes open after 40 to 45 days, and weaning begins when they are 10 weeks old. It is completed when they are 14 weeks old. Both sexes are sexually mature when they are about 10 months old.

SIMILAR SPECIES: In the Rockies, only the Abert's Squirrel (p. 238) is as large, but it has conspicuous ear tufts and only occurs in the southern Rockies, outside the Eastern Fox Squirrel's current range. The Red Squirrel (p. 240) is much smaller and redder. The Northern Flying Squirrel (p. 242) is smaller and grayer.

Abert's Squirrel

Sciurus aberti

Living in coniferous forests in the southernmost Rocky Mountains is the easily identified Abert's Squirrel. It has long ear tufts and a broadly plumed tail that are conspicuous even in dense forests. These ear tufts are so long that they bend in the breeze and sometimes become tipped with frost on chilly winter mornings. Because of these impressive ear tufts, the Abert's Squirrel is sometimes given the name "tassel-eared squirrel." Even if this squirrel were not bestowed with such handsome ears, it would still be considered one of the more beautiful squirrels because of its large, bushy tail, which is highlighted with white.

Abert's Squirrels build spherical tree nests out of twigs and leaves. They are accomplished architects, and some nests can measure almost 3 ft (1.1 m) in diameter. On occasion, these squirrels make nests in the matted clumps of pine twigs that occur naturally when a pine is infected with mistletoe. The squirrels hollow out these clumps and stuff them with soft leaves for bedding. Abert's Squirrels use their nests as a refuge during the day, and for sleeping at night. They do not hibernate in winter, but very cold winter weather may keep them in their nests.

Throughout the year, Abert's Squirrels find and store food. They feed primarily on the seeds and inner bark of ponderosa pine, pinyon nuts and an assortment of other vegetation. No food is stored in their nest; instead they prefer to bury their collections near their home. As is true for most squirrels, they sometimes forget where their caches are hidden, and the buried seeds may germinate. By forgetting cached food, all squirrels are helpful agents of forest propagation.

ALSO CALLED: Tassel-eared Squirrel.

DESCRIPTION: This rather large squirrel has conspicuous ear tufts, which are more than 1 in (2.5 cm) long in winter but are often shorter in summer. The tail is quite bushy and has white highlights. There are two basic color phases: the light phase is gray on the back, with a black stripe on each side of a white belly and a gray tail fringed with white; the dark phase squirrels are uniformly dark brown or black. Dark-phase individuals are predominant in northern Colorado, whereas in southern Colorado the light-phase squirrels are more common, many with a rather broad line of reddish hairs along the back.

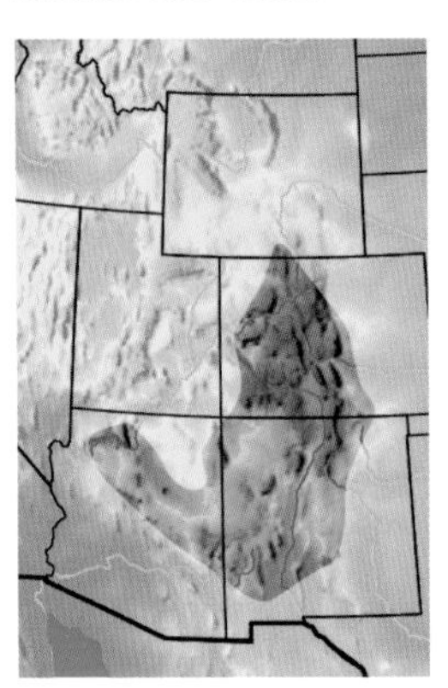

RANGE: Abert's Squirrels are found in pine-forested areas from southern Wyoming southward through Colorado, New Mexico and Arizona.

DID YOU KNOW?

There is a handsome subspecies of the Abert's Squirrel, called the Kaibab Squirrel, that has a totally white tail and lives only on the north rim of the Grand Canyon.

Total Length: 18–23 in (46–58 cm)
Tail Length: 7–10 in (18–25 cm)
Weight: 24–32 oz (680–910 g)

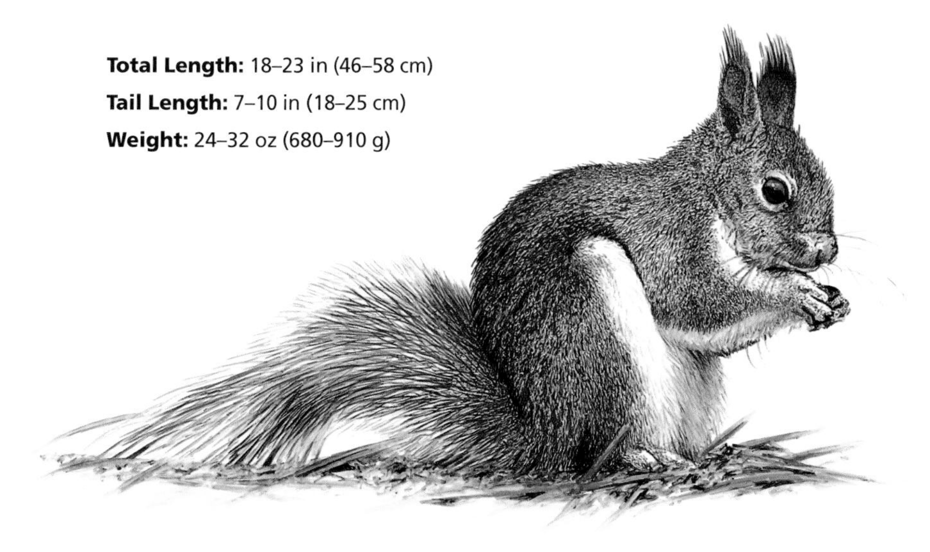

HABITAT: Abert's Squirrels are closely associated with ponderosa pine forests, usually at an elevation of 5000–8000 ft (1520–2440 m).

FOOD: Pinecones, the sweet inner bark of pine twigs, pine buds and a few fungi that grow on or around trees are the main foods for this squirrel.

DEN: The spherical nest, located high in the foliage of a ponderosa pine, is intricately constructed of pine needles, bark, sticks and grass, and it can sometimes reach almost 3 ft (1.1 m) in diameter. Abert's Squirrels occasionally build their nests among the tangled twigs that result from mistletoe infections.

YOUNG: Usually there is a single litter of three to four pups born in May or early June. They appear to be sexually mature following their first winter.

SIMILAR SPECIES: The Eastern Fox Squirrel (p. 236) lacks the ear tufts and has a more northern and eastern distribution. The Red Squirrel (p. 240) is much smaller and redder. The Northern Flying Squirrel (p. 242) is also smaller and is a sooty pewter color.

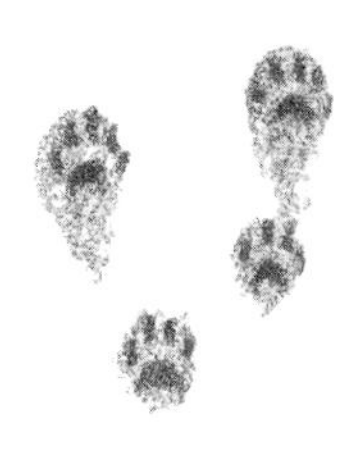

bounding trail

Red Squirrel

Tamiasciurus hudsonicus

Few squirrels have earned such a reputation for playfulness and agility as the Red Squirrel. This mammal is a well-known backyard and ravine inhabitant that often has a saucy regard for its human neighbors. Like a one-man-band, the Red Squirrel firmly scolds all intruders with shrill chatters, clucks and sputters, falsettos, tail flicking and feet stamping. Even when it is undisturbed, this chatterbox often chirps as it goes about its daily routine. A Red Squirrel is delightful to watch as it sits on a branch and scolds from a safe distance; it is very active and, by comparison, we are slow and awkward along the forest floor.

For this industrious squirrel, the daytime hours are devoted almost entirely to food gathering and storage. It urgently collects conifer cones, mushrooms, fruits and seeds in preparation for the winter months. The Red Squirrel remains active throughout the winter, except in severely cold weather. At temperatures below –13°F (–25°C) it stays warm, but awake, in its nest.

Because the Red Squirrel does not hibernate, it needs to store massive amounts of food in winter caches. These food caches, which in extreme cases can reach the size of a garage, are the secret to the Red Squirrel's winter success. Much of its efforts throughout the year are concentrated on filling these larders, and many biologists have speculated that the Red Squirrel's characteristically antagonistic disposition is a result of having to continuously protect its food stores.

By the end of winter, Red Squirrels are ready to mate. Their courtship involves daredevil leaps through the trees and chases over the forest floor. A female may mate repeatedly with more than one male. Between the months of April and June, two to seven pink, blind babies are born. The youngsters are playful and frequently challenge nuts or mushrooms to aggressive combat.

DESCRIPTION: The shiny, clove brown summer coat sometimes has a central reddish wash along the back. A black longitudinal line on each side separates the dorsal color from the grayish to white underparts. There is a white eye ring. The backs of the ears and the legs are rufous to yellowish. The longest tail hairs have a black subterminal band and buffy tip, which gives the tail a light fringe. The longer, softer winter fur tends to be bright to dusky rufous on the upperparts, with fewer buffy areas, and the head and belly tend to be grayer. The whiskers are black.

RANGE: The Red Squirrel occupies coniferous forests across most of Alaska and Canada. In the West, it extends south through the Rocky Mountains to southern New Mexico. In the East, it occurs south to Iowa and Virginia and through the Alleghenies.

DID YOU KNOW?

In fall, Red Squirrels nip conifer cones from the tops of trees, letting them fall to the ground. Once they have harvested enough cones, the squirrels descend, collect the bounty and store it, often in heaps, for winter usage.

Total Length: 11–14 in (28–36 cm)

Tail Length: 4¼–5¾ in (11–15 cm)

Weight: 6–11 oz (170–310 g)

HABITAT: Boreal coniferous forests and mixed forests are the major habitat, but towns with trees more than 40 years old also support Red Squirrels.

FOOD: The majority of the diet consists of seeds extracted from conifer cones. A midden is formed where discarded cone scales and centers pile up below a favored feeding perch. Flowers, birds, berries, mushrooms, eggs, mice, insects and even baby Snowshoe Hares or chipmunks may be eaten.

DEN: Tree cavities, witch's broom (created in conifers in response to mistletoe or fungal infections), logs and burrows may serve as den sites. The burrows or entrances are about 5 in (13 cm) in diameter, with an expanded cavity housing a nest ball that is 16 in (41 cm) across.

YOUNG: Northern populations bear just one litter a year. Peak breeding, in April and May, is associated with frenetic chases and multiple copulations lasting up to seven minutes each. After a 35- to 38-day gestation, a litter of usually four or five helpless young is born. The eyes open at four to five weeks, and the young are weaned when they are seven to eight weeks old. Red Squirrels are sexually mature by the following spring.

SIMILAR SPECIES: In the Rocky Mountains, only the Northern Flying Squirrel (p. 242) is a similar size, but it is a sooty pewter color. The Abert's Squirrel (p. 238) and the Eastern Fox Squirrel (p. 236) are much larger.

Northern Flying Squirrel

Glaucomys sabrinus

Like drifting leaves, Northern Flying Squirrels seem to float from tree to tree in forests throughout much of the Rocky Mountains. These arboreal performers are one of two species of flying squirrels in North America that are capable of distance gliding.

Although it lacks the ability of true flapping flight—bats are the only mammals to have mastered that skill—a flying squirrel's aerial travels are no less impressive, with extreme glides of up to 110 yd (100 m). Enabling the squirrels to "fly" are its gliding membranes—cape-like, furred skin extending down the length of the body from the forelegs to the hindlegs.

Before a glide, a squirrel identifies a target and maneuvers into the launch position: a head-down, tail-up orientation in the tree. Then, using its strong hindlegs, the squirrel propels itself into the air with its legs extended. Once airborne, it resembles a flying paper towel that can make rapid side-to-side maneuvers and tight downward spirals. Such control is accomplished by making minor adjustments to the orientation of the wrists and forelegs. On the ground and in trees, flying squirrels hop or leap, but the skin folds prevent them from running. They do not seem able to swim, either.

The call of the Northern Flying Squirrel is a loud *chuck chuck chuck*, which increases in pitch to a shrill falsetto when it is disturbed. Like other tree-dwelling squirrels, the Northern Flying Squirrel does not hibernate. On severely cold days, however, groups of 5 to 10 individuals can be found huddled in a nest to keep warm.

DESCRIPTION: Flying squirrels have a unique web or fold of skin that extends laterally to the level of the ankles and wrists and serves as the abbreviated "wings" with which the squirrels glide. They have large, dark, shiny eyes. The back is light brown, with hints of gray from the lead-colored hair bases. The feet are gray on top. The underparts are light gray to cinnamon precisely to the edge of the gliding membrane and edge of the tail. The tail is noticeably flattened from top to bottom, which adds to the buoyancy of the "flight" and helps the tail function as the rudder and elevators do on a plane.

HABITAT: Coniferous mountain forests, especially those with old trees for nesting, are prime flying squirrel habitat, but these rodents are sometimes found in aspen and cottonwood forests.

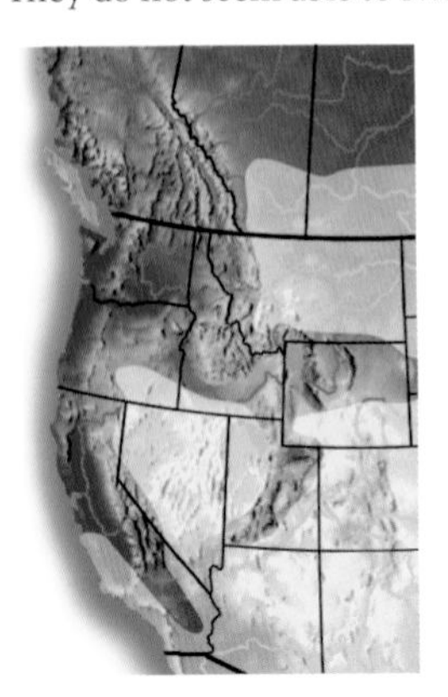

RANGE: This squirrel occurs in coniferous and mixed forests in eastern Alaska and across most of Canada. Its range extends south through the western mountains to California and Utah, around the Great Lakes and through the Appalachians.

DID YOU KNOW?

Northern Flying Squirrels may be just as common in an area as Red Squirrels, but their nocturnal activity patterns mean they are rarely seen. Flying squirrels may routinely visit bird feeders at night; they value the seeds as much as sparrows and finches do.

Total Length: 9¾–15 in (25–38 cm)

Tail Length: 4¼–7 in (11–18 cm)

Weight: 2⅝–6½ oz (74–180 g)

FOOD: The bulk of the food consists of lichens and fungi, but flying squirrels also eat buds, berries, some seeds, a few arthropods, bird eggs and nestlings, and the protein-rich, pollen-filled staminate cones of conifers. They cache cones and nuts.

DEN: Nests in tree cavities are lined with lichen and grass. Leaf nests, called dreys, are located in a tree fork close to the trunk. Twigs and strips of bark are used on the outside, with progressively finer materials used inside until the center consists of grasses and lichens. If the drey is for winter use, it is additionally insulated to a diameter of 16 in (41 cm).

YOUNG: Mating takes place between late March and the end of May. After a six-week gestation period, typically two to four young are born. They weigh about 3⁄16 oz (5.3 g) at birth. The eyes open after about 52 days. Ten days later they first leave the nest, and they are weaned when they are about 65 days old. Young squirrels first glide at three months; it takes them about a month to become skillful gliders. Flying squirrels do not become sexually mature until after their second winter.

SIMILAR SPECIES: No other mammal in the Rocky Mountains has the distinctive flight membranes of a flying squirrel. The Red Squirrel (p. 240) is reddish brown overall and is generally active during the day.

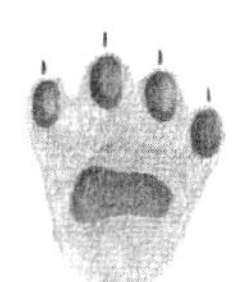

right foreprint

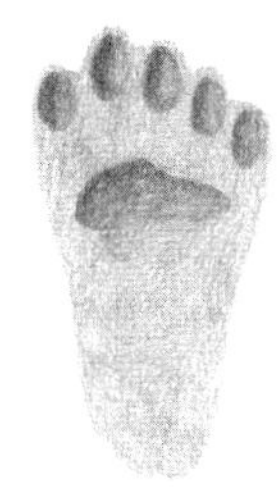

right hindprint

HARES & PIKAS

These rodent-like mammals are often called lagomorphs after the scientific name of the order, Lagomorpha, which means "hare-shaped." Rabbits, hares and pikas share a rodent's trademark, chisel-like upper incisors, and taxonomists once grouped the two orders together. Unlike rodents, however, lagomorphs have a second pair of upper incisors. Casual observers will never see these peg-like teeth; instead of being in the general tooth row, they lie immediately behind the first upper incisor pair.

Lagomorphs are strict vegetarians, but they have relatively inefficient, non-ruminant stomachs that have trouble digesting such a diet. To make the most of their meals, they defecate pellets of soft, green, partially digested material that they then reingest to obtain maximum nutrition. Some biologists believe that this process evolved as a protective mechanism that allows a lagomorph to quickly fill its stomach and then retreat to a hiding place to digest the meal in safety.

Hare Family (Leporidae)

Rabbits and hares are characterized by their long, upright ears, long jumping hindlegs and short, cottony tails. These timid animals are primarily nocturnal, and they spend most of the day resting in shallow depressions, called "forms." Rabbits build a maternity nest for their young, which are blind and naked at birth. Hares are born fully furred and with open eyes, and soon after birth they begin to feed on vegetation.

Pika Family (Ochotonidae)

Pikas are the most rodent-like of the lagomorphs, and, with their rounded ears and squat bodies, they look a lot like small guinea pigs. Their front and rear limbs are about the same length, so pikas scurry, rather than hop, through the rocky outcrops and talus of their home territories. Pikas are most active during the day, so they are often seen by hikers in mountain parks.

Pygmy Rabbit

Brachylagus idahoensis

Total Length: 10–12 in (25–30 cm)

Tail Length: 13⁄16–1¼ in (2.1–3.2 cm)

Weight: 10–15 oz (280–430 g)

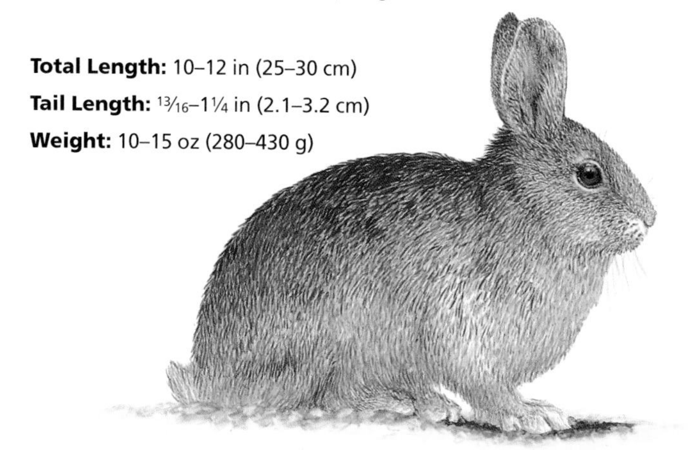

Unlike other rabbits, the Pygmy Rabbit excavates its own burrow in the hard soil of sagebrush flats. Pygmy Rabbits are not often encountered, but where they are found it is not uncommon to stir up several from the waist-high sagebrush. The frightened rabbits run to their burrow entrances, pause just outside to scan for danger, and then dart quickly into the burrow to make their escape.

DESCRIPTION: This tiny rabbit is usually pinkish-cinnamon. Sometimes the coat is dark gray to nearly black. There is a small white spot on the side of each nostril. The tail is entirely gray.

HABITAT: The Pygmy Rabbit is always found in association with dense stands of sage or rabbitbrush. It favors desert or semi-desert conditions where it can find earth soft enough for digging.

FOOD: The bitter leaves of the sagebrush make up the bulk of the diet. When available, grasses and other succulent vegetation are eaten.

DEN: This small rabbit digs burrows that are about 3 in (7.6 cm) in diameter. There are at least two entrances, usually located beneath large, dense sagebrush plants. A network of shallow trenches typically radiates out from an entrance, and the rabbits often crouch in a trench so that only their ears and eyes are visible.

YOUNG: Pygmy Rabbits mate in spring or early summer. After a gestation of 27 to 30 days, a litter of six naked, blind young are born between late May and early August.

SIMILAR SPECIES: The Pygmy Rabbit is so small that the only other rabbit it could be confused with is a juvenile Mountain Cottontail (p. 246), although Mountain Cottontails show white on the tail.

RANGE: The Pygmy Rabbit is primarily a Great Basin animal that ranges between the Rocky Mountains and Cascades in Montana, Idaho, Utah, Nevada, California, Oregon and Washington.

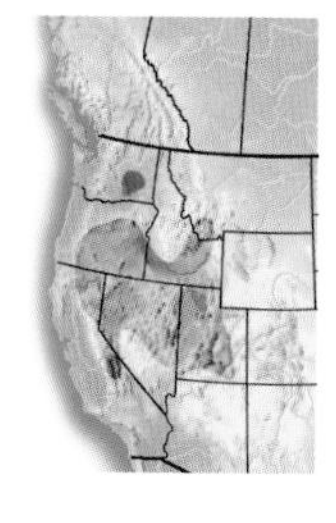

Mountain Cottontail

Sylvilagus nuttallii

When the sun lowers to meet the mountain horizon, flooding mid-summer evenings in golden light, Mountain Cottontails emerge from their daytime hideouts to graze on succulent vegetation. They can often be closely observed in foothill areas as they daintily nip at grasses, always just a short leap from dense bushes or a rocky shelter.

These rabbits fit the cuteness niche in the minds and hearts of most people: a cottontail's coal black eyes, rounded ears and soft features correspond to almost everyone's image of a wild bunny. Their plush-toy appearance, however, masks a tough nature: the Mountain Cottontail is quite capable of surviving in a harsh, unforgiving landscape full of predators.

Mountain Cottontails spend most of their days sitting quietly in dug-out depressions, called "forms," beneath impenetrable vegetation or under rocks, boards, abandoned machinery or buildings. These mid-sized herbivores have small home ranges that rarely exceed the size of a baseball field. Heavy rains greatly diminish cottontail activity, restricting them to their hideouts for the duration of the storm. Mountain Cottontails do not hibernate during winter, but they limit their movements to traditional trails that they can easily locate after a snowfall.

Thomas Nuttall is a wondrous choice to be immortalized in the name of this inquisitive and endearing mammal. Although Nuttall was primarily a botanist, he made significant contributions to all fields of natural history. He was also renowned for his absent-mindedness and misadventures. Many of his most famous gaffes occurred during his voyage across the continent to the Pacific on the Wyeth expedition in 1834. On several occasions he became lost, and should any animal provide a sense of comfort to one so misguided, it would surely be the cottontail that now bears his name.

ALSO CALLED: Nuttall's Cottontail.

DESCRIPTION: This rabbit has dark, grizzled, yellowish-gray upperparts and whitish underparts year-round. The tail is blackish above and white below. There is a rusty-orange patch on the nape of the neck, and the front and back edges of the ears are white. The ears are usually held erect when the rabbit runs.

HABITAT: A major habitat requirement is cover, whether it is brush, fractured

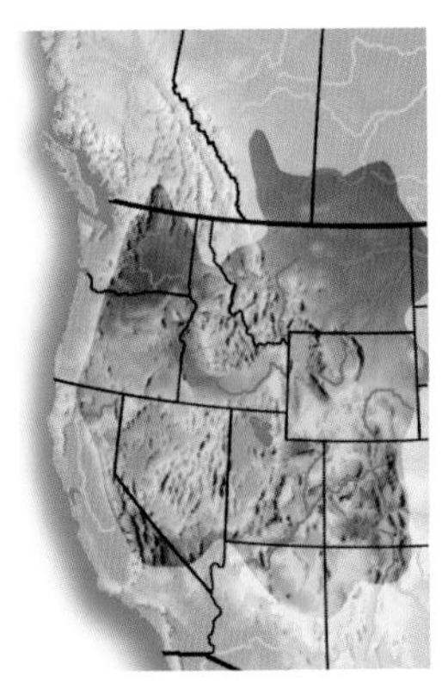

RANGE: The western limit of this rabbit's range closely matches the eastern slopes of the Cascades and Sierra Nevada, running north into the Okanagan of British Columbia. The eastern edge of the range runs south from southern Saskatchewan along the Montana-Dakota border to northern New Mexico.

DID YOU KNOW?

Rabbits and hares depend on intestinal bacteria to break down the cellulose in their diets. Because the bacterial products reenter the gut beyond the site of absorption, rabbits eat their pellets to run the material through the digestive tract a second time.

Total Length: 13–16 in (33–41 cm)

Tail Length: 1¼–2½ in (3.2–6.4 cm)

Weight: 1½–2¼ lb (680–1020 g)

rock outcrops or buildings. These rabbits like edge situations where trees meet meadows or where brushy areas meet agricultural land.

FOOD: Grasses and forbs are the primary foods, but in many areas they feed heavily on sagebrush and juniper berries.

DEN: There is no true den, but cottontails shelter in a form dug out beneath and among rocks, boards, buildings and the like. The young are born in a nest that is dug out by the female and lined with grass and fur. The doe arrives at the nest and lies over the top while the young nurse. The nest is essentially invisible, and a casual observer would never suspect that the female was nursing, or that a nest of babies lay beneath her.

YOUNG: Breeding begins in April, and after a 28- to 30-day gestation a litter of one to eight (usually four or five) young is born. The female is in estrus again and breeds within hours of giving birth, so there can be two litters a season. The young are born blind, hairless and with their eyes closed. They grow quickly and are weaned just before the birth of the subsequent litter.

SIMILAR SPECIES: The Desert Cottontail (p. 248) tends to have less hairy ears and a brownish patch on the nape of the neck. The Pygmy Rabbit (p. 245) is much smaller. Both the Snowshoe Hare (p. 250) and the White-tailed Jackrabbit (p. 252) are much larger and become white in winter.

Desert Cottontail

Sylvilagus audubonii

Old age is unknown to the Desert Cottontail. As an animal that lives near the bottom of the food web, each turn in life holds extreme risk. This cottontail does not lament its fate, but rather exploits all its opportunities for success. Even in the first year of life, reproduction affords some individuals the opportunity to contribute to the continuation of their kind. At the age of two, healthy females can have up to four litters between May and September, which is a necessity, because few, if any, reach age three.

During summer, Desert Cottontails are most active between sundown and dawn. Foraging in low light conditions probably gives the rabbits some measure of protection against predators. In winter, the rabbits reverse their habits and become more active during the day. When pursued, they tend to dodge this way and that until finally reaching shelter.

Desert Cottontails are quite unlike most other kinds of rabbits. They tend not to rest during the day in worn beds, but rely on other burrowing species for dens. The abandoned burrows of ground squirrels, prairie-dogs and even badgers are used as sites to hole-up during the day or inclement weather. Despite their affinity for subterranean resting places, these rabbits have been known to climb up on fallen and leaning trees to better view their surroundings.

Although these characteristics were largely unknown when the Desert Cottontail was initially described, the tribute of its scientific name to John James Audubon is most appropriate. Audubon is primarily known as an expert on birds, but he also made significant contributions to some of the first descriptive treatments of American mammals. More appropriate still, Audubon, like the Desert Cottontail, was an independent spirit and did things his way, including climbing up trees to get a better look at the world.

DESCRIPTION: The Desert Cottontail is a rusty grayish brown above. The sides are paler than the back and may have a yellowish wash. The throat and upper forelegs are rusty. The belly is white. The tail is dark above and white below.

HABITAT: This rabbit favors sagebrush-covered slopes, in association with fractured rock outcrops, and the brushy streamside areas that occur along the infrequent small watercourses in its arid territory.

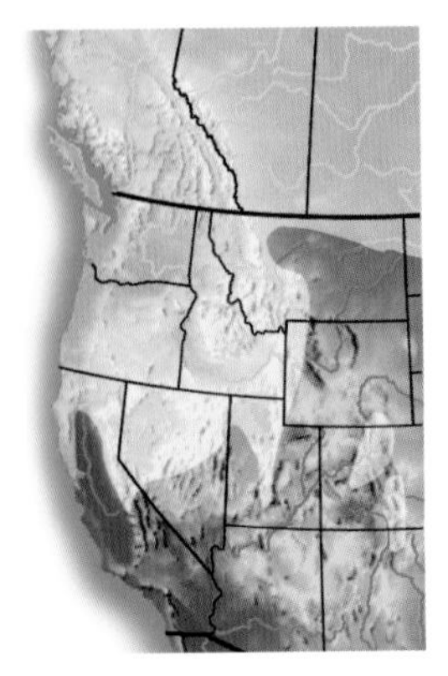

RANGE: Desert Cottontails range from western North Dakota, central Montana and northern California south into Mexico.

DID YOU KNOW?

Unlike other rabbits, the Desert Cottontail has special adaptations for life in arid environments. In particular, its kidneys are designed to resorb more water and concentrate salts to reduce water loss in the urine.

Total Length: 15–18 in (38–44 cm)

Tail Length: 1½–2⅜ in (3.9–5.8 cm)

Weight: 2–2½ lb (910–1130 g)

FOOD: The succulent tips of grasses, forbs and some sagebrush are preferred. It eats more brushy material in winter.

DEN: This cottontail may shelter in prairie-dog burrows, among shattered rocks, beneath buildings or in scrap lumber piles, but the young are generally born into a pear-shaped nest dug by the mother and lined with grass and fur.

YOUNG: After a 28- to 30-day gestation period, the first litter is born in May. Two to four litters in a season are possible because the female mates again within hours of giving birth. The blind, naked young grow rapidly. Their eyes open in 10 days, and the young leave the nest when they are just two weeks old. The newly emerged young are incredibly attractive with their large eyes, long ears and tiny, "powder-puff" tails. Unfortunately for many, night-hunting owls consider them a tasty meal. Spring born females may have a litter during their first summer, but they are not full-grown until they are six to nine months old.

SIMILAR SPECIES: The Mountain Cottontail (p. 246) has hairier ears with white edges and a rusty-orange patch on the nape of the neck. The Pygmy Rabbit (p. 245) is much smaller. Both the Snowshoe Hare (p. 250) and the White-tailed Jackrabbit (p. 252) are much larger and become white in winter.

Snowshoe Hare

Lepus americanus

As an animal highly adapted to withstand the most unforgiving aspects of the Rocky Mountain wilderness, the Snowshoe Hare possesses several fascinating adaptations for winter. As its name implies, for example, the Snowshoe Hare has very large hindfeet that enable it to cross areas of soft snow where other animals sink into the powder. This ability is usually a tremendous advantage for an animal that is preyed upon by so many different species of carnivores. Unfortunately, it is of minimal help against the equally big-footed Canada Lynx, a specialized hunter of the Snowshoe Hare.

It is quite well known that populations of lynx and hares fluctuate in nearly direct correlation with one another, but few people realize that other species are involved in the cycle. Recent studies have shown that as hares increase in number, they overgraze willow and alder in their habitat. These plants are their major source of food during the winter months. In response to overgrazing, the willows and alders produce a distasteful and toxic substance in their shoots that is related to turpentine. This substance protects the plants and initiates starvation in the hares. As the hares decline, so do the lynx. Once the plants recover their growth after a season or two, their shoots become edible again and the hare population increases.

In response to shortening day lengths at the onset of winter, Snowshoe Hares start changing into their white winter camouflage whether snow falls or not. The hares have no control over the timing of this transformation, and if the year's first snowfall is late, some individuals will lose their usual concealment, becoming visible from great distances—to naturalists and predators alike—as bright white balls in a brown world. The hares seem to be aware of this handicap, however, and they will often seek out any small patch of snow on which to squat.

DESCRIPTION: The summer coat is rusty brown above, with the crown of the head being darker and less reddish than the back. The nape of the neck is grayish brown and the ear tips are black. The chin, belly and lower surface of the tail are white. Adults have white feet; immatures have dark feet. In winter, the terminal portion of nearly all the body hair becomes white, but the hair bases and the underfur are lead gray to brownish. The ear tips remain black.

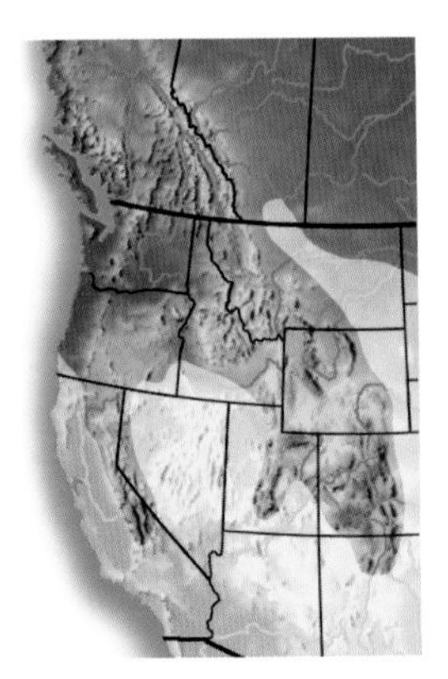

RANGE: The range of the Snowshoe Hare is associated with the boreal coniferous forests and mountain forests from northern Alaska and Labrador south to northern New Jersey in the East and to California and New Mexico in the West.

DID YOU KNOW?

Between their highs and lows, Snowshoe Hare densities can vary by a factor of 100. During highs, there may be five to six hares per acre (12 to 15 per hectare); after a population crashes, they may be uncommon over huge geographical areas for years.

Total Length: 15–21 in (38–53 cm)

Tail Length: 1⅞–2⅛ in (4.8–5.4 cm)

Weight: 2¼–3¼ lb (1–1.5 kg)

summer colors

HABITAT: Snowshoe Hares may be found almost everywhere there is forest or dense shrub in the Rocky Mountains.

FOOD: In summer, a wide variety of grasses, forbs and brush may be consumed. In winter, mostly the buds, twigs and bark of willows and alders are eaten. Hares will occasionally eat carrion.

DEN: Snowshoe Hares do not keep a customary den, but they sometimes enter hollow logs or the burrows of other animals, or run beneath buildings.

YOUNG: Breeding activity begins in March and continues throughout August. After a gestation period of 35 to 37 days, one to seven (usually three or four) young are born under cover, but often not in an established form or nest. The female breeds again within hours of their birth, and she may have as many as three litters in a season. The young hares are born with fur and with their eyes open. They can hop within a day, are active within a week and are feeding on grassy vegetation within 10 days. In five months they are full grown.

SIMILAR SPECIES: The White-tailed Jackrabbit (p. 252) has longer ears and a slightly longer tail, and its winter underfur is creamy white. The Black-tailed Jackrabbit (p. 254) has longer ears and a black uppersurface on its tail, and it doesn't turn white in winter. The Mountain Cottontail (p. 246) is generally smaller, has an orangish or rusty nape of the neck, and its ears are generally uniform in color or have white edges.

White-tailed Jackrabbit

Lepus townsendii

The White-tailed Jackrabbit is the largest and most commonly encountered hare through much of the Rocky Mountains. A creature of open country, the White-tailed Jackrabbit is most frequently encountered either by day as it bursts from a hiding place with its ears erect and its tail extended, bounding out of danger with ease, or at night in the flash of car headlights. These lean sprinters are customarily solitary when they are out foraging, but in winter up to 50 may gather in one place, often where food is abundant. These gatherings usually occur at night, which is when jackrabbits are most active.

Like most herbivores, the White-tailed Jackrabbit is drawn to salt, which, unfortunately, it often licks from roads. This need for salt, together with its preference for traveling on solid surfaces, may partly explain the large numbers of road-killed jackrabbits encountered on some low-elevation highways.

Their protruding eyes and straight limbs give White-tailed Jackrabbits a "meaner" look than the other rabbits, but these features are innocent adaptations for detecting and avoiding predators. Known for their raw speed, long-legged jackrabbits can outdistance most land-based predators in an all-out run. Ambush appears to be the most effective method of catching a jackrabbit, but the open country in which these animals live gives little opportunity for cover-seeking predators. Golden Eagles have moderate successes in attacks from overhead, occasionally inflicting mortal injury to the backs of these speedsters.

White-tailed Jackrabbits can sometimes be found using the same rest areas day after day, so if you spook a jackrabbit from its hideout during a walk, return cautiously the next day to look for it in the same area. Quite often, these large hares can be found using exactly the same forms, and they can be surprisingly approachable if your movements are slow and unthreatening.

DESCRIPTION: Like the Snowshoe Hare, this jackrabbit changes color seasonally. In summer, the upperparts of this large hare are light grayish brown and the belly is nearly white. By mid-November, the entire coat is white, except for the grayish forehead and the black ear tips. It has a fairly long, white tail that sometimes bears a grayish band on the upper surface. The tail is held rigidly behind the animal as it runs.

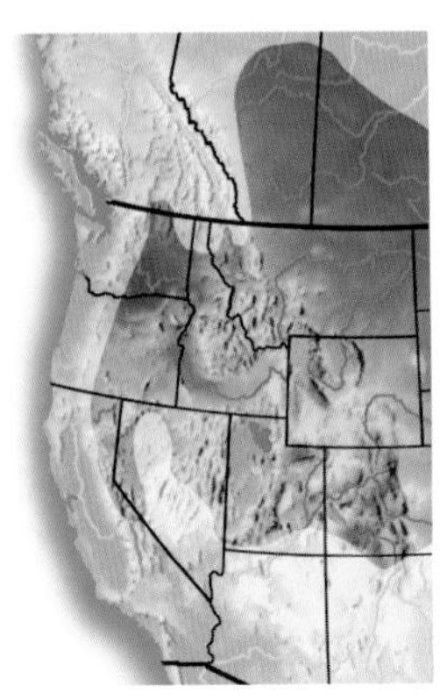

RANGE: The White-tailed Jackrabbit seems to be expanding its range northward, perhaps as a result of land clearing. It is currently found from eastern Washington east to southern Manitoba and south to central California and eastern Kansas.

DID YOU KNOW?

John Kirk Townsend, honored in the scientific name townsendii, *is a largely underrated naturalist. Best known for his work on the 1834 Wyeth expedition, he likely encountered many White-tailed Jackrabbits while crossing the northern plains.*

Total Length: 21–25 in (53–64 cm)

Tail Length: 2¾–4¼ in (7–11 cm)

Weight: 6½–12 lb (2.9–5.4 kg)

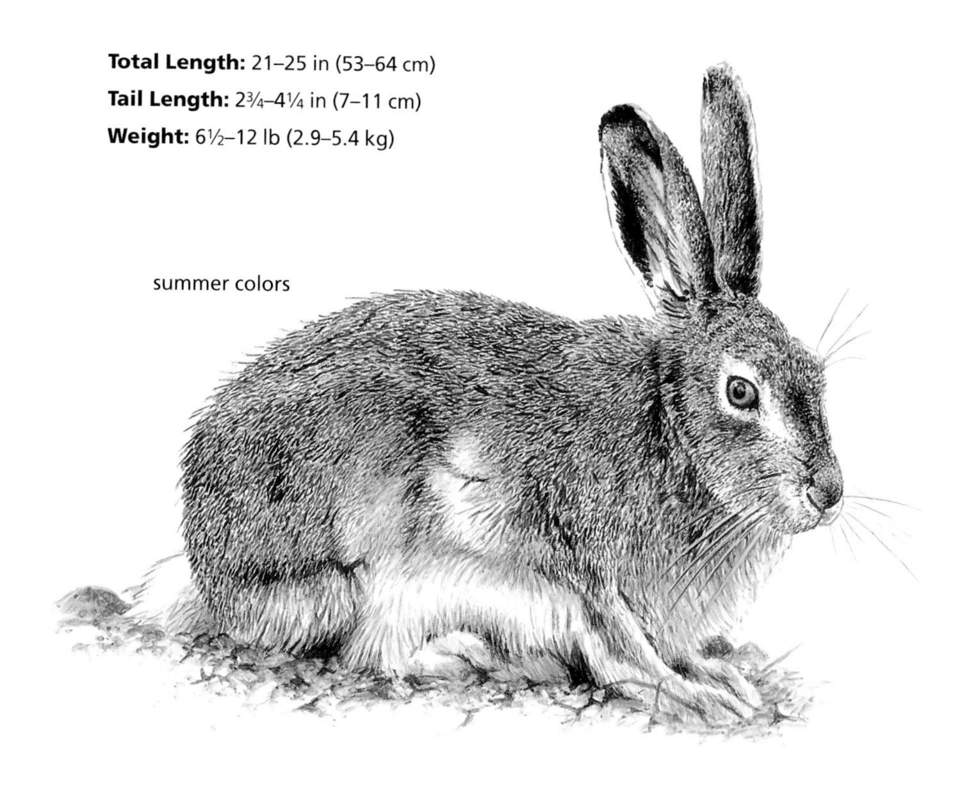

summer colors

HABITAT: This hare is a creature of open areas. It will enter open woodlands to seek shelter in winter, but it avoids dense, timbered stands.

FOOD: Grasses and forbs are the most commonly eaten plants, but jackrabbits also enjoy alfalfa and clover in agricultural regions. More shrubs and weedy plants appear in the winter diet. Like all hares, they eat their pellets to run the bacteria and bacterial breakdown products from their cecum through the digestive system a second time, allowing greater absorption of the nutrients.

DEN: There is no den, but a shallow form beside a rock or beneath sagebrush serves as a daytime shelter. In winter, jackrabbits may dig depressions or short burrows as shelters in snowdrifts.

YOUNG: One to nine (usually three or four) young are born in a shallow depression after a 40-day gestation. The fully furred newborns have open eyes and soon disperse, meeting their mother to nurse only once or twice a day. By two weeks old they are eating some green vegetation; at five to six weeks they are weaned, often just before the birth of the next litter. Jackrabbits reach adult weight in three to four months.

SIMILAR SPECIES: In winter, the Snowshoe Hare (p. 250) has lead gray, not creamy white, hair bases, and it is nearly always associated with treed or brushy areas. The Black-tailed Jackrabbit (p. 254) has longer ears and a black uppersurface on its tail, and it doesn't turn white in winter.

Black-tailed Jackrabbit

Lepus californicus

You can forget the tail; this hare's ears are its trademark. Black-tipped ears that extend far above this hare's head are a feature that, more than any other, identifies this large, open-country mammal.

The three hares of the Rocky Mountains demonstrate a relationship between the size of their ears and the latitude in which they live: the Black-tailed Jackrabbit, the most southerly hare, has the longest ears; the White-tailed Jackrabbit has intermediate ears; and the most northerly Snowshoe Hare has the shortest ears. Huge ears, such as those of the Black-tailed Jackrabbit, help cool this hare, which occupies hot, arid areas. At the other extreme, the Snowshoe Hare's smaller ears prevent precious energy from escaping during cold winters. Ear size may also contribute to hearing ability—sound moves better through cold air than it does through warm air, and hares that live in warmer climates may have evolved larger ears to better listen for approaching predators.

The Coyote is the major predator of the Black-tailed Jackrabbit, and in some areas the density of Coyotes correlates with Black-tailed Jackrabbit numbers. In turn, the density of Black-tailed Jackrabbits was directly proportional to summer precipitation.

During the day, Black-tailed Jackrabbits rest in a form scraped out beneath a sagebrush bush or beside some other type of cover. These hares generally lie quietly, depending upon their camouflage for protection. With the approach of a predator, however, jackrabbits rocket from their shelter and quickly attain speeds of up to 35 mph (56 km/h). They may attempt to elude danger by changing direction abruptly and using Olympian leaps up to 6 ft (1.8 m) high and 20 ft (6.1 m) long.

DESCRIPTION: This gray to grayish-brown hare has extremely large, long, black-tipped ears. The belly is white to buffy white. There is a black mid-dorsal stripe on the tail that runs up onto the back. These hares never turn white in winter and there is only one annual molt.

HABITAT: The Black-tailed Jackrabbit occupies nearly all habitats within its range, except high mountain forests. It prefers valley bottoms, and it is often found in irrigated fields in intermontane valleys, although it is no stranger to barren ridges devoid of trees.

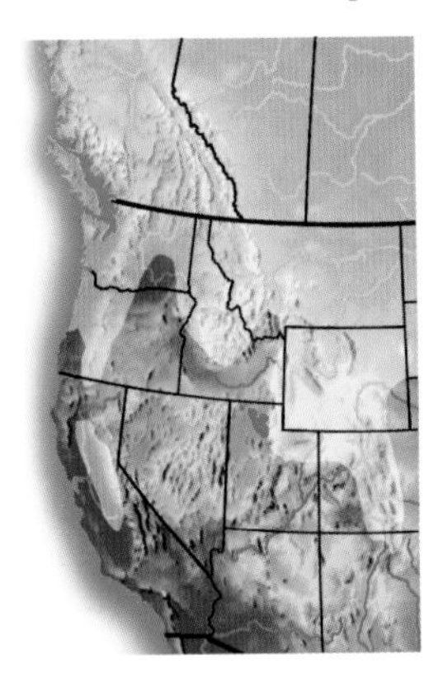

RANGE: Black-tailed Jackrabbits currently occupy the southwestern quarter of North America, but they appear to be slowly expanding their range northward. In some places they may be displacing White-tailed Jackrabbits.

DID YOU KNOW?

The name "jackrabbit" is a shortened and more refined version of "jackass rabbit." These mammals were so-named in recognition of their large, donkey-like ears.

Total Length: 20–24 in (51–61 cm)
Tail Length: 2¾–3¾ in (7–9.5 cm)
Weight: 5–10 lb (2.3–4.5 kg)

FOOD: These jackrabbits feed on a wide variety of both herbaceous and woody vegetation, with a larger proportion of shrubby material being consumed in winter. Studies have revealed that 30 Black-tailed Jackrabbits eat about as much as a cow. They may have peculiar eating habits: one plant may be eaten in its entirety, while neighboring plants of the same species are ignored.

DEN: There is no den, but the hare spends most of the day crouched in a form it scratches out of the ground.

YOUNG: After a gestation of 41 to 47 days, a litter of one to eight (usually two to four) young is born. The newborns are fully furred and have their eyes open. Instead of staying in a common nest, they are distributed more than 300 ft (91 m) apart, and the female nurses each one separately throughout the night. The young are weaned at six to seven weeks. At 10 weeks they are 90 percent of their adult size. Within hours after birth, the female becomes attractive to a male, although she flees at his approach. A vigorous chase ensues that may cover a few miles. Ultimately, the male seizes the female by the nape of her neck and mates with her. Females may have up to four litters in a season.

SIMILAR SPECIES: The White-tailed Jackrabbit (p. 252) has slightly shorter ears, turns white in winter and has a grayish uppersurface on its tail. The Snowshoe Hare (p. 250) has shorter ears, turns white in winter and has an all-white tail year-round.

American Pika

Ochotona princeps

Inhabiting a confusing landscape of boulders nestled in a rocky mountain cradle, the American Pika is one of the Rockies' top cuteness contenders. This relative of the rabbit scurries among the talus rocks of a landslide as it makes its way between its feeding areas and shelter. Frequently, an American Pika returns from its nearby foraging areas with vegetation clippings held crossways in its mouth. The bundle is sometimes half as large as the pika itself, and the vegetation is carried back toward the den, where it is accumulated in piles on and around the rocks to be dried for winter use.

Pikas are extremely vocal animals that are often heard before they are seen. The proper pronunciation of their name is *pee-ka,* which is somewhat reminiscent of their high-pitched voices—they give out tricycle horn bleats whenever they see something out of the ordinary. Their voices are the first, and often best, clues of pika activity, because these animals are so difficult to distinguish in their bouldered habitats—when a pika is momentarily glimpsed from afar, one is never quite sure whether it is a genuine sighting or just a pika-sized rock. To the patient naturalist intent on pika observations, however, viewing can be intimate and rewarding, because these animals often permit a close approach. When you see a pika escape into a crevice beneath the rocks, sit quietly and wait—soon it will come again, seemingly oblivious to your unobtrusive presence.

In winter, pikas dig snow tunnels as far as 100 yd (91 m) out from their rock shelters to collect and eat plants. The talus slopes that are their homes often receive great quantities of snow, which helps insulate the animals from the mountain winters. Rarely venturing into the chill of the open air, pikas tend to remain beneath the snow, sometimes feeding upon the grass they so meticulously gathered and dried earlier in the year.

DESCRIPTION: This gray to tawny-gray, chunky, soft-looking mammal has large, rounded ears and beady black eyes. The whiskers are long. There is no external tail. The front and rear legs are nearly equal in length, so pikas run instead of hopping.

HABITAT: Pikas generally occupy accumulations of broken rock, known as talus slopes, in the mountains, although they have rarely been found among the jumbled logs swept down by avalanches.

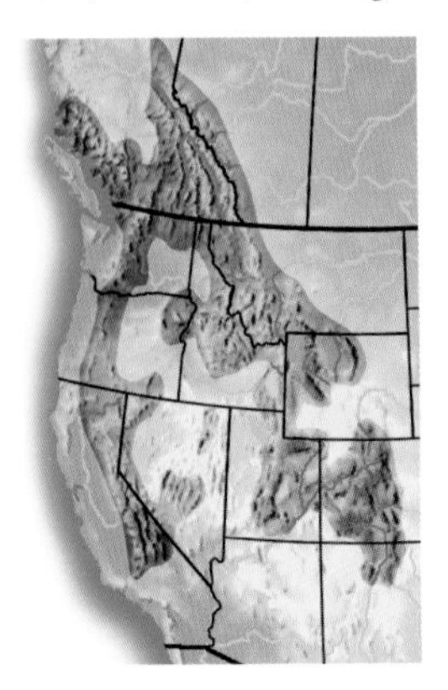

RANGE: Pikas occur in the mountains from west-central Alberta and eastern British Columbia south to California, Utah and New Mexico.

DID YOU KNOW?

Pikas seem to throw their voices, so although they may often be heard, they can be difficult to locate. This ventriloquist-like ability is a great advantage to an individual pika that wants to warn its neighbors without revealing itself to a potential predator.

Total Length: 7–7¾ in (18–20 cm)

Tail Length: There is no external tail.

Weight: 6⅛–8⅛ oz (174–230 g)

FOOD: The diet includes a wide variety of plants that are found in the vicinity of this animal's rocky shelter. Broad-leafed plants, grasses and sedges are all clipped and consumed.

DEN: Pikas build grass-lined nests, in which the young are born, beneath the rocks of their home.

YOUNG: Mating occurs in spring, and after a 30-day gestation period a litter of two to five (usually three) young is born. The newborns are furry, weigh ¼–5⁄16 oz (7.1–8.9 g) and have closed eyes. The eyes open after 10 days. The young are weaned when they are 30 days old and two-thirds grown. Pikas are sexually mature after their first winters. There is sometimes a second summer breeding period.

SIMILAR SPECIES: The only other gray mammal of comparable size that might occupy the same rocky slopes as the American Pika is the Bushy-tailed Woodrat (p. 170), but woodrats have long, bushy tails.

foreprint

hindprint

BATS

Only three groups of vertebrates have achieved self-powered flight: bats, birds and the now-extinct reptilian pterosaurs. In an evolutionary sense, bats are a very successful group of mammals. Worldwide, nearly a quarter of all mammalian species are bats, and they are second only to rodents in both diversity of species and number of individuals.

Unlike the feathered wings of a bird, a bat's wings consist of double layers of skin stretched across the modified bones of the fingers and back to the legs. A small bone, called the calcar, juts backward from the foot to help support the tail membrane, which stretches between the tail and each leg. The calcar is said to be keeled when there is a small projection of skin from its side.

Bats generate lift by pushing their wings against the air's resistance. This method of flight is less efficient than the airfoil lift provided by bird or airplane wings—bats have to have large wing surface areas for their body size—but it allows bats to fly slower and gives them more maneuverability. Slower flight is a real advantage when they are trying to catch insects or hovering in front of a flower.

Bats have excellent vision, but their nocturnal habits have led to an increased dependence on their hearing—most people are acquainted with the ability of many bat species to navigate or capture prey in the dark using echolocation. The tragus, a slender lobe that projects from the inner base of many bats' ears, is thought to help in determining an echo's direction. Each species of bat in the Rockies echolocates at different frequencies, so a person equipped with a bat detector—these things actually exist—can identify the species of a bat from its ultrasonic nighttime clicks.

Big Free-Tailed Bat

Free-tailed Bat Family (Molossidae)

The free-tailed bats are so named because they have naked tails that extend beyond the edge of the membrane that stretches between their hindlegs. Most free-tailed bats occur in warm tropical parts of the world, but two species may be found in the southern parts of the Rocky Mountains. Free-tailed bats are sometimes called mastiff bats, because their snub noses and wrinkled faces resemble those of mastiff dogs.

Little Brown Bat

Evening Bat Family (Vespertilionidae)

The majority of the bats that occur in the Rocky Mountains belong to the evening bat family. True to their name, most members of this family are active in the evening, and often again before dawn, when they typically feed on flying insects. A few species migrate to warmer regions for winter, but most hibernate in caves or abandoned buildings and mines.

Brazilian Free-tailed Bat

Tadarida brasiliensis

Total Length: 3½–4½ in (8.9–11 cm)

Tail Length: 1¼–1¾ in (3.2–4.5 cm)

Forearm: 1½–1¾ in (3.8–4.5 cm)

Weight: 5⁄16–7⁄16 oz (8.9–12 g)

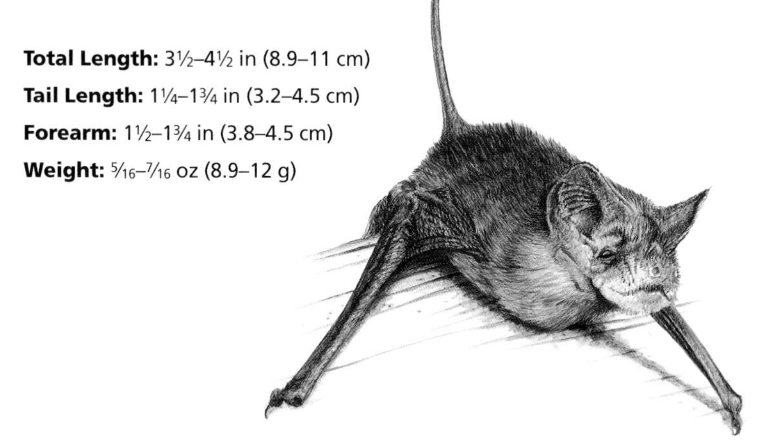

More than any other bats, Brazilian Free-tailed Bats display the massiveness of their population numbers. In the Carlsbad Caverns of New Mexico, for instance, the collective wingbeats of millions of free-tailed bats leaving their daytime roosts sound like the roar of a whitewater river, and the rising column can be seen at great distances. Colonies in the southern Rockies do not reach the tremendous numbers that are seen elsewhere, but at low elevations within their limited mountain range they are still fairly common.

DESCRIPTION: This small free-tailed bat is dark brown to grayish on the back and lighter on the underside. The ears are separated at the base. The upper lip is wrinkled. There is little membrane connecting the hindlegs and tail, thus the name "free-tailed."

HABITAT: In the southern Rockies, these bats are mostly found in pinyon-juniper woodlands.

FOOD: These bats forage primarily on such night-flying insects as beetles, ants and moths.

DEN: Individuals are known to roost in buildings, but it is in caves that their numbers become great. Large populations can carpet the walls and ceilings of caves and mines, with as many as 250 individuals per square foot (2700 individuals per square meter).

YOUNG: Females ovulate for a short period of time in March, and mating is spread out over a five-week overlapping period. One or two young are born in June. At birth, the young are two-thirds the length of their mothers; within three weeks they equal her in mass.

SIMILAR SPECIES: The Big Free-tailed Bat (p. 260), the only other free-tailed bat normally found within this region, is much larger and less common.

RANGE: The Brazilian Free-tailed Bat is one of the most wide-ranging mammals in the New World. In North America, it occurs from southern Oregon to North Carolina and south through Mexico.

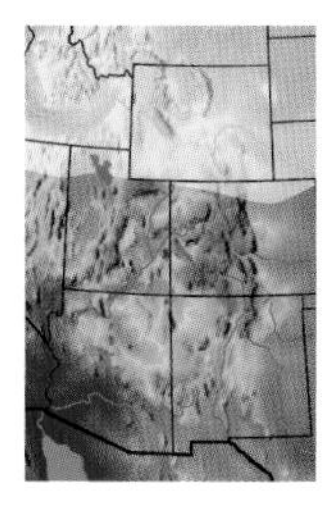

Big Free-tailed Bat

Nyctinomops macrotis

Total Length: 5⅛–5⅝ in (13–14 cm)

Tail Length: 1¾–2⅛ in (4.5–5.4 cm)

Forearm: 2¼–2½ in (5.7–6.4 cm)

Weight: 1¼–1½ oz (35–43 g)

The Big Free-tailed Bat is a giant among Rocky Mountain bats, but even at that, it is little larger than a sparrow. The size of this bat's wings is what gives it the illusion of being larger. Unlike birds, bats have large heads (with teeth), solid bones, heavy limbs and fleshy tails, all of which require relatively larger wings to get bats into flight. Their large wings enable bats to fly at lower speeds, however, and with far more maneuverability than birds, which is a great advantage when catching slow-flying insects.

DESCRIPTION: This large bat has a wingspan of about 18 in (46 cm). Its fur is reddish brown to black, but the base of each hair is white. Most of the tail length (about 1 in [2.5 cm]) is free of the membrane joining the hindlegs. The ears are jointed at the base, and they are long enough to extend beyond the nose if they are pushed forward.

RANGE: This free-tailed bat occurs from Utah and Colorado south through southeastern California and western Texas.

HABITAT: These bats are most common in rocky areas, particularly cliffs and crevices.

FOOD: Like many nocturnal bats, they feed primarily on moths, but they will opportunistically take whatever other edible insects can be caught.

DEN: Roost sites are often in rock crevices, cliffs and caves, but these bats have also been found in buildings and hollow trees.

YOUNG: Only a single young is born each year, in late spring or early summer.

SIMILAR SPECIES: The only other free-tailed bat in the Rockies is the Brazilian Free-tailed Bat (p. 259), which is much smaller and more common.

Fringed Bat

Myotis thysanodes

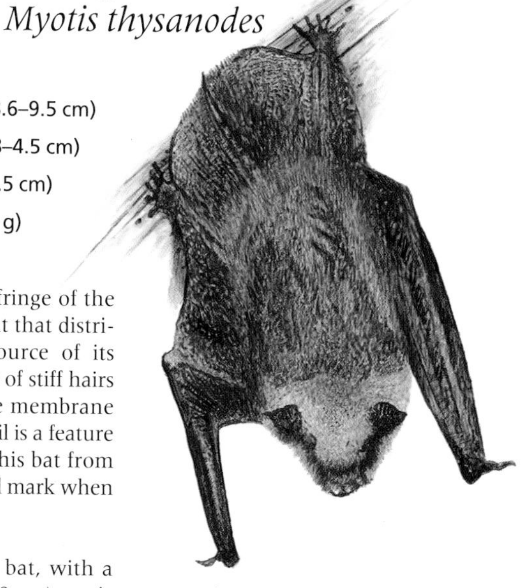

Total Length: $3\frac{3}{8}$–$3\frac{3}{4}$ in (8.6–9.5 cm)

Tail Length: $1\frac{1}{2}$–$1\frac{3}{4}$ in (3.8–4.5 cm)

Forearm: $1\frac{5}{8}$–$1\frac{3}{4}$ in (4.1–4.5 cm)

Weight: $\frac{3}{16}$–$\frac{5}{16}$ oz (5.3–8.9 g)

This bat exists on the fringe of the Rocky Mountains, but that distribution is not the source of its name. A conspicuous fringe of stiff hairs along the outer edge of the membrane between the hindlegs and tail is a feature that usually distinguishes this bat from others, but it is a useless field mark when the bat is in flight.

DESCRIPTION: This large bat, with a wingspan of up to 12 in (30 cm), typically has pale brown fur that is darker on the back than on the undersides. The long, blackish ears contrast strongly with the color of the back and would extend well past the nose if pushed forward. The most unique characteristic of this bat is the fringe of small stiff hairs on the outer edge of the tail membrane. The calcar is keeled.

HABITAT: Throughout most of the Rocky Mountains, this bat is most often encountered along streams in grasslands and ponderosa pine forests. It occurs mainly at low elevations in the northern parts of its range.

FOOD: Moths, flies, beetles, lacewings and crickets are fairly commonly eaten, typically shortly after sunset. The presence of flightless insects in the diet has led to the speculation that these bats may glean some insects from foliage.

DEN: These bats roost in caves, mines and buildings. Up to several hundred Fringed Bats will cluster in maternal roosts in summer. Maternal colonies contain only females and that year's young.

YOUNG: One or uncommonly two young are born in June or early July. The young bats reach adult size by three weeks, at which time they are capable of limited flight.

SIMILAR SPECIES: The Long-eared Bat (p. 262) is slightly smaller, has larger ears and lacks the stiff hairs on the tail membrane. The Long-legged Bat (p. 269) also lacks the fringe on the tail membrane, and its underwings are furred to the elbow and knee.

RANGE: The Fringed Bat is found from southern British Columbia south through the western states into Mexico.

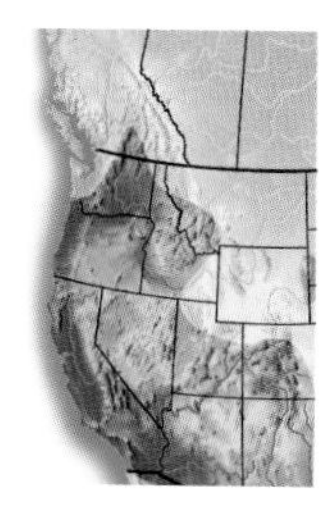

Long-eared Bat

Myotis evotis

Total Length: 3¼–4¼ in (8.3–11 cm)

Tail Length: 1⅜–1⅞ in (3.5–4.8 cm)

Forearm: 1½–1⅝ in (3.8–4.1 cm)

Weight: ⅛–5⁄16 oz (3.5–8.9 g)

The nightly, dramatic bat sagas taking place in the summer skies are largely unknown to humans. In apparent silence, Rocky Mountain bats navigate and locate prey by producing ultrasonic pulses (up to five times higher in pitch than our ears can detect) and listening for the echoes of these sounds as they bounce off objects. Bats with large ears tend to be insectivorous, and the aptly named Long-eared Bat should be well-equipped for insect hunting.

DESCRIPTION: The wingspan of this medium-sized bat is about 11 in (28 cm). The upperparts are light brown to buffy yellow. The undersides are lighter. Its black, naked ears are ¾–1 in (1.9–2.5 cm) long. The tragus is long and narrow. The wings are mainly naked, and only the lower fifth of the tail membrane is furred. The calcar is keeled.

RANGE: The Long-eared Bat is found from southern British Columbia east to extreme southern Saskatchewan and south to northwestern New Mexico and Baja California.

HABITAT: This bat occurs in forested areas adjacent to rocky outcrops or badland landscapes. It occasionally occupies buildings, mines and caves.

FOOD: Feeding peaks at dusk and just before dawn. Moths, flies and beetles are the primary prey.

DEN: Both sexes of this mainly solitary bat hibernate in caves and mines in winter. In spring, groups of up to 30 females gather in nursery colonies in tree cavities, under loose bark, in old buildings, under bridges or in loose roof shingles. Males typically roost in caves and mines in summer.

YOUNG: Mating takes place in fall, before hibernation begins, but fertilization is delayed until spring. In June or early July, after a gestation of about 40 days, a female bears one young. Twins are uncommon. The young mature quickly and are able to fly on their own in four weeks.

SIMILAR SPECIES: All the mouse-eared bats (*Myotis* spp.) are generally indistinguishable in flight.

Northern Bat

Myotis septentrionalis

The Northern Bat tends to roost in natural cavities and under peeling bark on old trees during the warmer months in the northern parts of the Rockies. Some people are concerned that it is therefore vulnerable to forestry operations, which often select older trees for harvesting. Only a few caves in the Rockies are known to host hibernating Northern Bats.

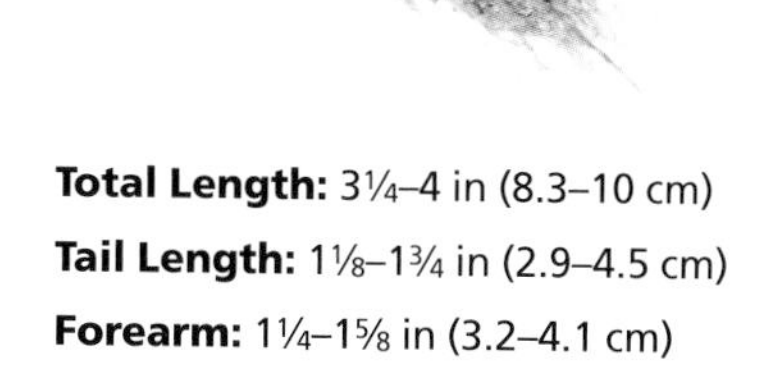

Total Length: 3¼–4 in (8.3–10 cm)

Tail Length: 1⅛–1¾ in (2.9–4.5 cm)

Forearm: 1¼–1⅝ in (3.2–4.1 cm)

Weight: ⅛–5⁄16 oz (3.5–8.9 g)

DESCRIPTION: This mid- to dark brown bat has a wingspan of 9–10 in (23–25 cm). It has somewhat lighter underparts. The tips of its hairs are lighter than the bases, giving a sheen to the fur. The tragus is long and narrow. The calcar is not keeled.

HABITAT: The Northern Bat, which primarily occurs in forested and sometimes brushy areas, prefers to be close to waterbodies.

FOOD: This bat primarily feeds at dusk and again just before dawn. It catches flying insects, including moths, flies and beetles.

DEN: In September and October, this mainly solitary bat seeks out caves and mines in which to hibernate. The males continue to roost in caves and mines all year. The females form nursery colonies in spring. These colonies are usually located in tree cavities, under loose bark on trees, in old buildings, under bridges or in loose shingles on rooftops, and they may be small or contain up to 30 females.

YOUNG: These bats mate in fall, but fertilization is delayed until spring, so the single young is not born until June or early July, after a gestation of about 40 days. The young are able to fly in about four weeks.

SIMILAR SPECIES: Mouse-eared bats (*Myotis* spp.) are generally indistinguishable from one another in the field. Identification of the species requires a good key, precise measurements and careful attention to detail.

RANGE: Northern Bats are found from eastern British Columbia east to Newfoundland and south to Nebraska, Arkansas, western Georgia and Virginia.

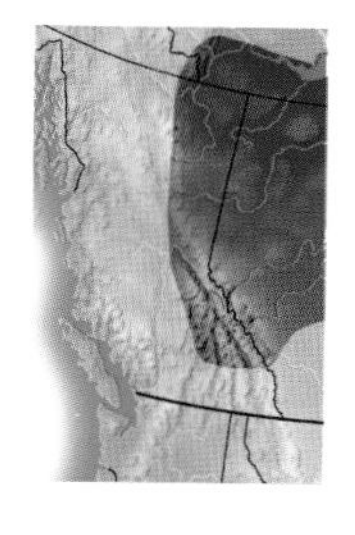

California Bat

Myotis californicus

Total Length: 3–3¾ in (7.6–9.5 cm)

Tail Length: 1¼–1⅝ in (3.2–4.1 cm)

Forearm: 1¼–1⅜ in (3.2–3.5 cm)

Weight: ⅛–3⁄16 oz (3.5–5.3 g)

California Bats emerge shortly after nightfall, and for a few minutes in the dying daylight, they can be followed as they fly erratically through the sky. As if surfing on invisible waves in the air, fluttering California Bats rise and dive at variable speeds in the pursuit of unseen prey. These activities appear disorganized and random, but they are actually deliberate and calculated.

DESCRIPTION: This bat is small and yellowish brown. Its wingspan is about 9 in (23 cm). Its feet are tiny, and it has a keeled calcar.

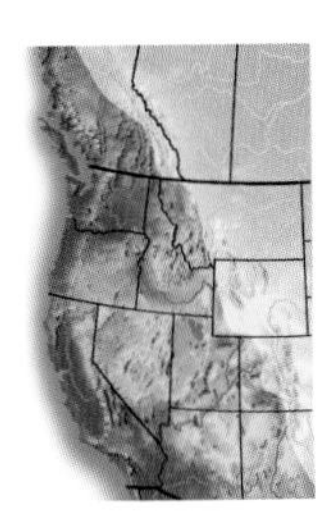

RANGE: The California Bat, truly a western bat, is found in coastal regions from southern Alaska south to California and Mexico. It ranges east into Montana, Colorado, New Mexico and Texas.

HABITAT: This bat has been found in lowland montane valleys, generally west of the main Rocky Mountain range. It can be found roosting in rock crevices, mines and buildings and under bridges and loose tree bark. It forages over montane forests and arid grasslands.

FOOD: During the night, California Bats forage opportunistically in areas that concentrate night-flying insects, such as cliffs and poplar groves, or over water for emerging adult caddisflies and mayflies. Additionally they can be observed in tree canopies feeding on moths, beetles and flies. Generally they fly 6–10 ft (1.8–3 m) above the ground when foraging.

DEN: California Bats are not too selective of their night roosts, and they have been found in buildings and natural structures. Their day roosts are typically found in rock crevices, but also in mine shafts, tree cavities, buildings and bridges.

YOUNG: A single young is born in late June to early July.

SIMILAR SPECIES: The Western Small-footed Bat (p. 265), which is more widespread through the Rockies, is almost indistinguishable in flight, but in the hand its dark face mask is distinctive.

Western Small-footed Bat

Myotis ciliolabrum

Total Length: 3–3½ in (7.6–8.9 cm)

Tail Length: 1³⁄₁₆–1¾ in (3–4.5 cm)

Forearm: 1⅛–1⅜ in (2.9–3.5 cm)

Weight: ⅛–¼ oz (3.5–7.1 g)

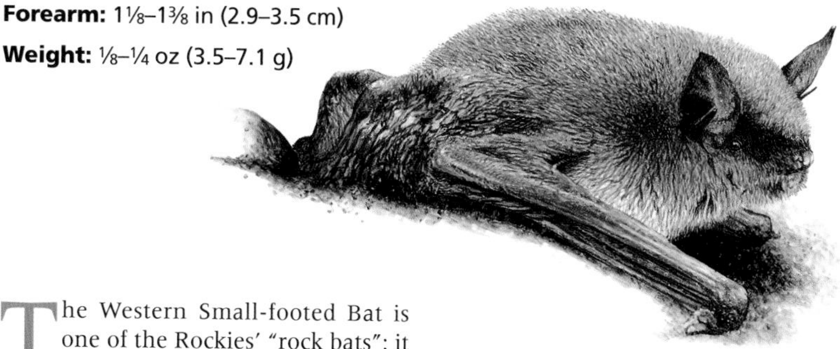

The Western Small-footed Bat is one of the Rockies' "rock bats": it occupies daytime roosts in such rocky habitats as badlands, cliffs and talus slopes. Summer colonies may contain dozens to hundreds of individuals, but the entire Rocky Mountain population possibly weighs less than just one big Grizzly Bear. Unlike the Grizzly, these bats receive little attention from mountain park visitors. In their nightly excursions, Western Small-footed Bats take to the skies cloaked in quietness.

DESCRIPTION: The glossy fur of this bat is yellowish brown to gray or even coppery brown above, and its undersides are almost white. The flight membranes and ears are black. The tail membrane is dark brown. The wingspan is 8–10 in (20–25 cm). Some fur may be found on both the undersurface of the wing and the upper surface of the tail membrane. Across its face, from ear to ear, is a dark brown or black "mask." True to its name, this bat has noticeably small feet. The calcar is strongly keeled.

HABITAT: The Western Small-footed Bat occupies drier habitats than any other bat in the Rockies. It prefers arid rocky or grassland regions, especially riverbanks, ridges and outcroppings with abundant rocks for roosting.

FOOD: Like most Rocky Mountain bats, the Western Small-footed Bat primarily eats flying insects, including moths, flies, bugs and beetles.

DEN: In summer, this bat roosts in trees, buildings or rock crevices. It hibernates in caves or mines in winter. Nursery colonies occur in bank crevices, under bridges or under the shingles of old buildings.

YOUNG: In small nursery colonies, one young per female is born from late May to early June.

SIMILAR SPECIES: In dim light, you cannot see the Western Small-footed Bat's "mask"; a technical key to the mammals of this region is needed to distinguish between the myotis bats.

RANGE: The Western Small-footed Bat is found from southern British Columbia east to southwestern Saskatchewan and south through most of the western U.S.

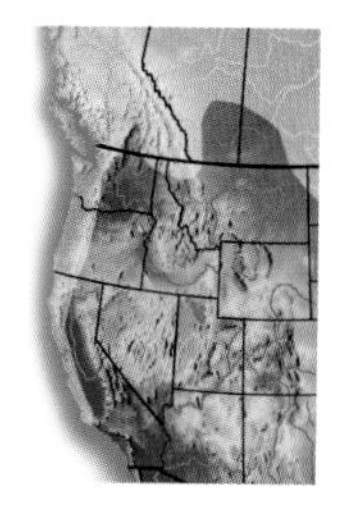

Little Brown Bat

Myotis lucifugus

On nearly every warm, calm summer night, the skies of the Rocky Mountains are filled with marvelously complex screams and shrills. Unfortunately for people interested in the world of bats, these magnificent vocalizations occur at frequencies higher than our ears can detect. The most common of these nighttime screamers, and quite likely the first bat most people will encounter, is the Little Brown Bat.

Once the cold days of late August and September arrive, Little Brown Bats begin to migrate to areas where they will spend the winter. Prior to entering hibernation, they mate. The young are not born until late June or early July, however, because fertilization of the egg is delayed until spring.

While it is not known where all of the West's Little Brown Bats spend the winter, thousands of them travel to caves in the mountains and foothills. Large wintering populations are known to occur in large caves, and cave adventurers are advised to take special care not to disturb these hibernating animals. Slight disturbances and subtle shifts in temperature can awaken the bats, suggesting to them that spring has arrived. Unfortunately, any bat flying out of a cave during the winter months is sure to die from exposure to the low temperatures outside.

DESCRIPTION: As its name suggests, this bat is little and brown. Its coloration ranges from light to dark brown on the back, with somewhat paler undersides. The tips of the hairs are glossy, giving this bat a coppery appearance. The wing and tail membranes are mainly unfurred, although fur may appear around the edges. The calcar of this bat is long and unkeeled. The tragus, which is nearly straight, is half the length of the ear.

HABITAT: Little Brown Bats are the most frequently encountered bats in much of North America. They are at home almost anywhere; you may find them in buildings, attics and roof crevices, under loose bark on trees or under bridges. Wherever these bats are roosting, waterbodies are sure to be nearby—the bats need a place to drink and a large supply of insects for their nightly foragings.

FOOD: Little Brown Bats feed exclusively on night-flying insects. In the evening, these bats leave their day roosts and swoop down to the nearest water source to snatch a drink on the

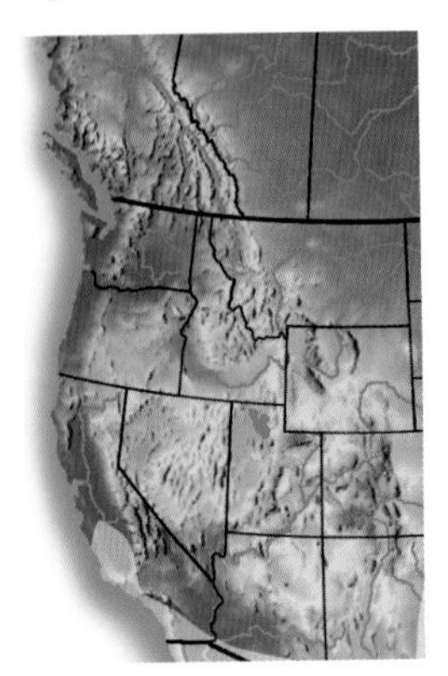

RANGE: This widespread bat ranges from central Alaska to Newfoundland and south to northern Florida, southeastern California and central Mexico, except for much of the southern Great Plains.

DID YOU KNOW?

An individual Little Brown Bat can consume 900 insects an hour during its nighttime forays. A typical colony may eat 100 lb (45 kg) of insects a year.

Total Length: 2⅜–4 in (6–10 cm)
Tail Length: 1–2⅛ in (2.5–5.4 cm)
Forearm: 1⅜–1⅝ in (3.5–4.1 cm)
Weight: 3⁄16–5⁄16 oz (5.3–8.9 g)

wing. Their foraging for insects can last for up to five hours. Later, the bats take a rest in night roosts (a different place from their day roosts). Another short feeding period occurs just before dawn, after which the bats return to their day roosts.

DEN: These bats may roost alone, in small groups or in colonies of more than 1000 individuals. A loose shingle, an open attic or a hollow tree are all suitable roosts for a Little Brown Bat. By June, pregnant females form nursery colonies in a protected location. In winter, some bats may stay and hibernate in large numbers in caves and old mines, but most are believed to migrate to warmer climates.

YOUNG: Mating occurs either in late fall or in the hibernation colonies. Fertilization is delayed until the females ovulate in spring, at which time they gather in nursery colonies. In June, one young is born to a female after about 50 to 60 days of gestation. The young are blind and hairless at birth, but their development is rapid and their eyes open in about three days. After one month, the young are on their own.

SIMILAR SPECIES: All the mouse-eared bats (*Myotis* spp.) are essentially impossible to identify in flight. Even in hand, one needs a technical key, which, through the examination of measurements or characteristics, allows determination of the species.

Yuma Bat

Myotis yumanensis

Total Length: 3–3⅝ in (7.6–9.2 cm)

Tail Length: 1¼–1¾ in (3.2–4.5 cm)

Forearm: 1¼–1½ in (3.2–3.8 cm)

Weight: ⅛–3/16 oz (3.5–5.3 g)

Like most bats, Yuma Bats spend much of the summer days hanging comfortably in hot roosts, shifting slightly as temperatures rise and fall. They rest, relax and snuggle against one another until the moon rises and draws them outside. With nightfall, Yuma Bats fly out over the nearest wetland, snapping up rising invertebrates in the cool night air. Often, their stomachs are full within 15 minutes and their foraging is finished for the night. They then return to their night roost, where they digest the night's offering before foraging again just before dawn.

DESCRIPTION: The medium-sized Yuma Bat has brown to black fur on the back, with a lighter color on the underside. The ears are long enough to extend to the nose when pushed forward. The tragus is blunt and only about half the length of the ear. The wingspan is about 9 in (23 cm). The calcar is not keeled.

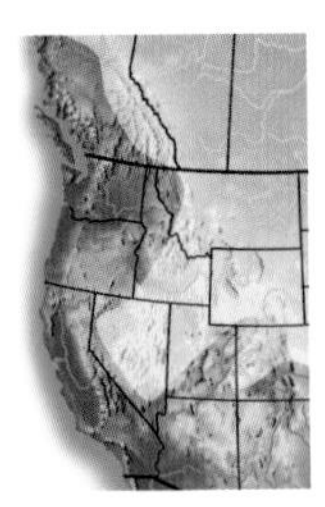

RANGE: The Yuma Bat is found from west-central British Columbia south to California and Mexico and east to Colorado and western Texas.

HABITAT: This bat tends to occur in grassland shrub areas of the Rockies, particularly in the south. It usually forages over lakes and streams.

FOOD: Much of the Yuma Bat's diet seems to consist of aquatic invertebrates, such as adult caddisflies, mayflies and midges.

DEN: Yuma Bats typically roost and form their maternal colonies in buildings, trees and caves and under south-facing siding and shingles. These structures must be within foraging distance of a source of water.

YOUNG: As in many types of bats, mating occurs during fall, with the sperm stored within the female until fertilization in the spring. A single young is usually born in June or July.

SIMILAR SPECIES: About the only way to distinguish the Yuma Bat from the Little Brown Bat (p. 266) externally is to notice its lack of the glossy or burnished tips on the dorsal fur, which makes its color slightly duller.

Long-legged Bat

Myotis volans

Total Length: 3⅜–4 in (8.6–10 cm)

Tail Length: 1⅜–2⅛ in (3.5–5.4 cm)

Forearm: 1⅜–1¾ in (3.5–4.5 cm)

Weight: 3/16–⅜ oz (5.3–11 g)

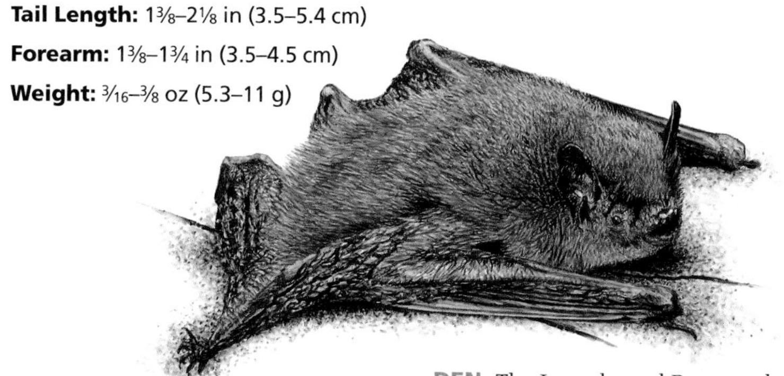

The leg bones of this bat are more than 5/8 in (1.6 cm) long, a characteristic responsible for its common name. Unfortunately, the Long-legged Bat's leg bones are only fractionally longer than those of the very similar Western Small-footed Bat, so they are not a good distinguishing characteristic. Noticeable differences do occur, however, in habitat selections. The Long-legged Bat always lives near water sources for their foraging opportunities.

DESCRIPTION: Although this bat is the heaviest of the "little brown bats," it is heavier by an almost imperceptible amount. The wingspan is 10–11 in (25–28 cm). The fur can be uniformly light brown to reddish to dark chocolate brown, but it is mainly dark brown. There is a well-defined keel on the calcar. The underwing is usually furred out to a line connecting the elbow and knee.

HABITAT: This bat lives primarily in coniferous forests that are near waterbodies. It may forage along the sides of mountain lakes.

FOOD: The diet is primarily composed of moths, flies, bugs and beetles.

DEN: The Long-legged Bat spends winter hibernating in caves or mines. In summer, it roosts in trees, buildings or rock crevices. Nursery colonies are located in bank crevices, under bridges or under south-facing shingles on old buildings.

YOUNG: Mating occurs in fall. Fertilization is delayed until spring, and the young are born in July or August, in large nursery colonies. One young per female is born. The young mature quickly, flying on their own in about four weeks. The longest recorded lifespan for this species is 21 years.

SIMILAR SPECIES: The slightly longer legs of the Long-legged Bat are not noticeable on a flying bat in dim light. To distinguish between the different myotis bats, you will need a technical key.

RANGE: The Long-legged Bat ranges from northwestern British Columbia southeast to western North Dakota and south through most of the western U.S.

Hoary Bat

Lasiurus cinereus

The Hoary Bat is one of the largest Rocky Mountain bats, with a wingspan of 15–16 in (38–41 cm), but it weighs less than the smallest mountain chipmunk. It flies later into the night than any other bat in the region, and once the last of the daylight has drained from the western horizon, the Hoary Bat courses low over wetlands, lakes and rivers in conifer country. It may not be as acrobatic in its foraging flights as the smaller myotis bats, but no one who has ever witnessed a Hoary Bat in flight would ever complain about its impressive aerial accomplishments.

The large size of the Hoary Bat is often enough to identify it, but the light wrist spots, which are sometimes visible, will confirm the identification. Many of the Hoary Bat's long hairs have brown bases and white tips, giving the animal an overall frosted appearance, hence its name. While attractive, this coloration makes the Hoary Bat very difficult to notice when it roosts in a tree—it looks very similar to dried leaves and lichens.

Hoary Bats, as well as other tree-dwelling bats, have recently been the focus of scientific study to determine the importance of old roost trees in bat ecology. These bats are quite complex animals: while old trees may well be important, water quality and the availability of hatching insects in wetlands may be equally significant.

The few records from the northern Rockies suggest that female Hoary Bats may migrate quite far north. The males, it is thought, migrate only as far as the northern U.S., where they likely court and mate with the females. While the males may remain at these sites for the summer, some impregnated females seem to push farther north, where the young are born.

DESCRIPTION: The large Hoary Bat has light brown to grayish fur, and the white hair tips give it a heavily frosted appearance overall. Its throat and shoulders are buffy yellow or toffee colored. Its wingspan is 15–16 in (38–41 cm). The ears are short, rounded and furred, but the edges of the ears are naked and black. The tragus is blunt and triangular. The upper surfaces of the feet and tail membrane are completely furred. The calcar is modestly keeled.

HABITAT: The Hoary Bat is often found near open grassy areas in coniferous and deciduous forests or over lakes.

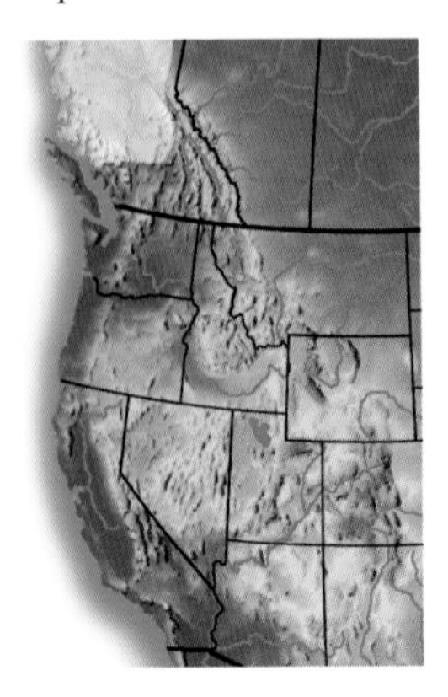

RANGE: From north-central Canada, the Hoary Bat ranges south through most of southern Canada and almost all of the lower U.S.

DID YOU KNOW?

The Hoary Bat is the most widespread species of bat in North America, and it is the only "terrestrial" mammal native to the Hawaiian Islands.

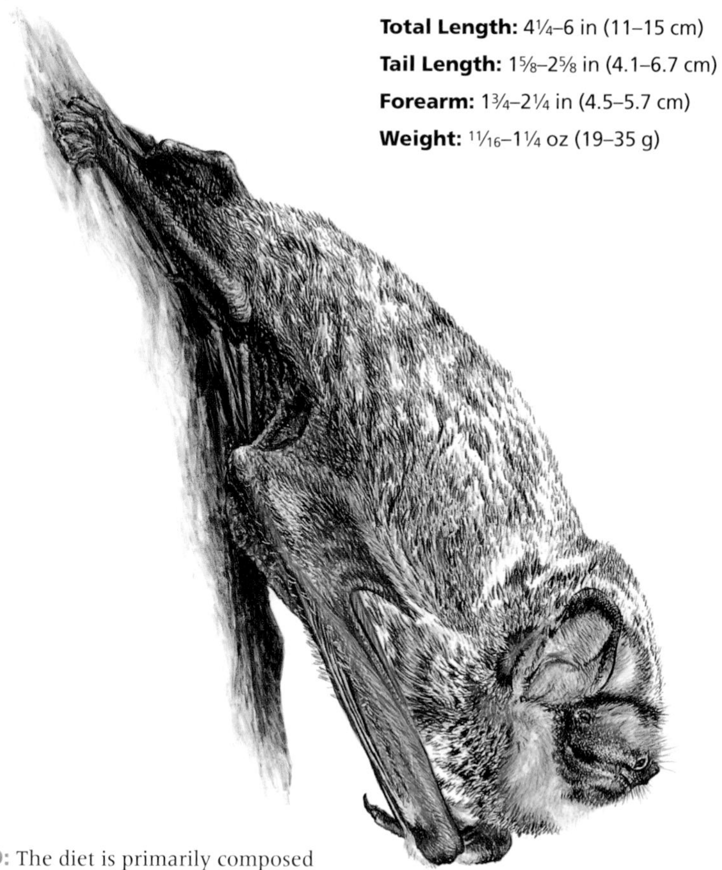

Total Length: 4¼–6 in (11–15 cm)
Tail Length: 1⅝–2⅝ in (4.1–6.7 cm)
Forearm: 1¾–2¼ in (4.5–5.7 cm)
Weight: 11/16–1¼ oz (19–35 g)

FOOD: The diet is primarily composed of moths, plant hoppers, flies and beetles, and may include many agricultural pests when this bat forages near farmlands. It sometimes alights on vegetation to pick off insects. Feeding activity does not peak until well after dusk.

DEN: This migratory bat usually returns to the central and northern Rockies in May. During summer, it roosts alone in the shade of foliage, with an open space beneath the roost so that it can drop into flight. Beginning in August or September, it migrates south, sometimes in large flocks.

YOUNG: Hoary Bats mate in fall, but the young are not born until late May or June because fertilization is delayed until the female ovulates in spring. Gestation lasts about 90 days, and a female, which has four mammae, usually bears two young. She places the first young on her back while she delivers the next. Before they are able to fly, young bats roost in trees and nurse between their mother's nighttime foraging flights.

SIMILAR SPECIES: The Silver-haired Bat (p. 272) is black with silver-tipped hairs and is slightly smaller. The Big Brown Bat (p. 273) is almost as large and does not have a frosted appearance.

Silver-haired Bat

Lasionycteris noctivagans

Total Length: 3⅝–4¼ in (9.2–11 cm)

Tail Length: 1⅜–2 in (3.5–5.1 cm)

Forearm: 1½–1¾ in (3.8–4.5 cm)

Weight: ¼–⅝ oz (7.1–18 g)

Silver-haired Bats, which are really quite attractive, could do much to soften anti-bat feelings if only they could be observed with regularity. They fly slowly and leisurely, fairly low to the ground, and they don't seem to be disturbed by the presence of an inquisitive human. Like most of the Rocky Mountains' bats, however, they are active at night and so are infrequently seen.

DESCRIPTION: The fur is nearly black, with long, white-tipped hairs on the back giving it a frosty appearance. The naked ears and tragus are short, rounded and black. The wingspan is 11–12 in (28–30 cm). A light covering of fur may be seen over the entire surface of the tail membrane.

RANGE: This bat is found across the southern half of Canada, including the southeastern coast of Alaska, south through most of the U.S.

HABITAT: Forests are the primary habitat, but this bat can easily adapt to parks, cities and farmlands.

FOOD: This bat feeds mainly on moths and flies. It has two peak feeding times, at dusk and just before dawn, and it forages over standing water or in open areas near water.

DEN: The summer roosts are usually in tree cavities, under loose bark or in old buildings. In winter, these bats may hibernate in caves, mines or old buildings. Females form nursery colonies in summer.

YOUNG: Breeding takes place in fall or during a break in hibernation, but fertilization is delayed until the female ovulates in spring. In early summer, after a gestation of about two months, one or two young are born to each female. If a young bat falls from its mother, the female locates it by listening for its high-pitched squeaks, and it may be able to climb back up to find her.

SIMILAR SPECIES: The Silver-haired Bat's white-tipped, black hairs are unique among Rocky Mountain bats. The Hoary Bat (p. 270) has light brown to grayish fur. The Big Brown Bat (p. 273) has mainly brown, glossy fur.

Big Brown Bat

Eptesicus fuscus

Total Length: 3⅝–5½ in (9.2–14 cm)

Tail Length: ⅞–2⅜ in (2.2–6 cm)

Forearm: 1⅝–2⅛ in (4.1–5.4 cm)

Weight: 7/16–1 oz (12–28 g)

The Big Brown Bat is not overly abundant anywhere, but its habit of roosting and occasionally hibernating in human structures makes it a more commonly encountered bat. It is also the only bat that may be seen on warm winter nights, because it occasionally takes such opportunities to change hibernating sites. The relative frequency of Big Brown Bat sightings doesn't save it from the anonymity that plagues most bats, however, because the "big" in its name is relative—this sparrow-sized bat still looks awfully small against a dark night sky.

DESCRIPTION: This big bat is mainly brown, with lighter undersides, and its fur appears glossy or oily. On average, a female is larger than a male. The face, ears and flight membranes are black and mainly unfurred. The blunt tragus is about half as long as the ear. The calcar is usually keeled.

HABITAT: This large bat easily adapts to parks, cities and farmlands. In the wild, it typically inhabits forests.

FOOD: A fast flier, the Big Brown Bat feeds mainly on beetles and plant hoppers, rarely moths or flies. Near farmlands, it feeds heavily on agricultural pests. Foraging usually occurs at heights of no more than 30 ft (9.1 m), and the two peak feeding periods are at dusk and just before dawn.

DEN: In summer, this bat usually roosts in tree cavities, under loose bark or in buildings. It spends winter hibernating in caves, mines or old buildings. Nursery colonies are found in protected areas, such as tree cavities, large crevices or old buildings.

YOUNG: These bats breed in fall or during a wakeful period in winter, but fertilization is delayed until the female ovulates in spring. A female gives birth to one or two young in early summer, after about a two-month gestation.

SIMILAR SPECIES: The Big Brown Bat is hard to distinguish from other large bats, but the Hoary Bat (p. 270) has frosted brown or gray fur, and the Silver-haired Bat (p. 272) has frosted black fur. The myotis bats (pp. 261–69) are all smaller.

RANGE: This bat occurs across most of British Columbia and northern Alberta to southeastern Manitoba, and south through the lower U.S.

Spotted Bat

Euderma maculatum

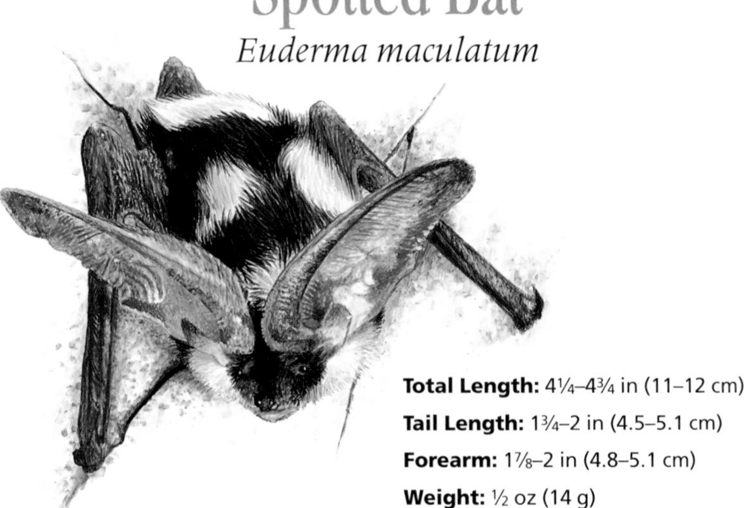

Total Length: 4¼–4¾ in (11–12 cm)

Tail Length: 1¾–2 in (4.5–5.1 cm)

Forearm: 1⅞–2 in (4.8–5.1 cm)

Weight: ½ oz (14 g)

The Spotted Bat is an exhibitionist among a guild of committed introverts. One glance at a Spotted Bat instantly reveals that it is simply no ordinary bat. The long, pink ears and the three huge, white spots that adorn its back are sufficiently distinctive for this bat to stand out in a crowd, but its flare is not only restricted to its appearance. While feeding, the Spotted Bat gives loud, high-pitched, metallic squeaks that are easily heard by humans. Because most bats vocalize beyond our hearing, it is unusually pleasing to listen to the aerial drama of the rare Spotted Bat.

DESCRIPTION: The back is primarily black. There is a large white spot on each shoulder and another on the rump. The belly is whitish. The long, pinkish to light tan ears project forward in flight but are folded back when the bat hangs up. The wingspan is about 12 in (30 cm).

HABITAT: These bats are found in highland ponderosa pine regions in early summer. They descend to lower-elevation deserts in August.

FOOD: Spotted Bats appear to be specialized predators on noctuid moths, a large and diverse family of night-flying insects. A few beetles have also been found in their stomachs.

DEN: In summer, Spotted Bats seem to roost primarily in rock cracks and crevices on cliffs and in caves.

YOUNG: Usually one young is born in early June (later in more northern areas). Even at a young age the ears are large, but the white spots on the back are absent on newborns.

SIMILAR SPECIES: The exceptionally large ears and large white spots make Spotted Bats distinctive among Rocky Mountain bats.

RANGE: This bat occurs from eastern Washington, southern Idaho and southern Montana south into central Mexico.

Pallid Bat

Antrozous pallidus

Total Length: 3¾–5⅜ in (9.5–14 cm)

Tail Length: 1⅜–2 in (3.5–5.1 cm)

Forearm: 1⅞–2⅜ in (4.8–6 cm)

Weight: 9⁄16–1¼ oz (16–35 g)

It might seem absurd that after millions of years of flight specialization, bats would be found foraging on the ground, but such is the case with Pallid Bats. These bats are still committed fliers, but they often land to take insects, other invertebrates and small vertebrates from the ground and vegetation.

While bats have few predators in the night skies, on the ground they are vulnerable to many threats. The light dorsal color of the Pallid Bat might be a protective adaptation that helps it blend in with the pale sands of its typically desert habitat.

DESCRIPTION: The back is pale yellow. The underparts are creamy or almost white. Individual hairs are always darker at the tip than at the base, which is a reverse of the typical situation for bats. The broad, tan ears are extremely long, and if pushed forward they may extend past the muzzle. The median edge of the ears is not folded. The wingspan is about 15 in (38 cm).

HABITAT: The Pallid Bat is typically associated with rocky outcrops near open dry areas, but occasionally it is found in coniferous forests in the southern Rockies.

FOOD: Insects are the main food, but some small vertebrates, such as lizards, have also been reported. It may incidentally eat some fruits and seeds.

DEN: Pallid Bats gather in night roosts following foraging. These sites are generally in caves, overhangs or buildings. Their day roosts are nearby, typically in buildings or rock crevices.

YOUNG: These bats mate from October through December and occasionally into February. The sperm is stored in the female's reproductive tract until ovulation in spring. Young are born in May and June, and twins are common.

SIMILAR SPECIES: The Townsend's Big-eared Bat (p. 276) is somewhat smaller and has lumps on its nose.

RANGE: The Pallid Bat ranges generally west of the Rockies from British Columbia to Baja California, extending east into Wyoming, Colorado and western Texas, and south to central Mexico.

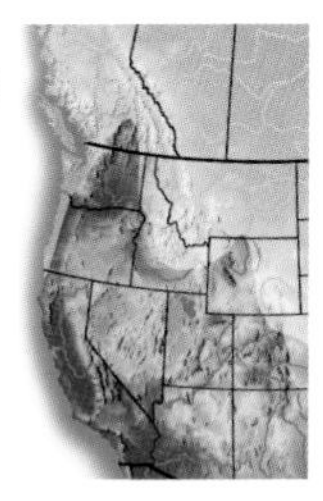

Townsend's Big-eared Bat

Plecotus townsendii

We all know that in Dumbo, Walt Disney went far beyond the realm of possibility in suggesting that this playful pachyderm could soar through the skies using its ears. Had the Disney creators searched for a more realistic character for this fable, they just might have turned to the Townsend's Big-eared Bat. Naturally, the oversized ears of this bat serve no more of a function in flight than do the ears of any other mammal, but for sheer size and believability, few animals have ears to match.

As humans, we tend to perceive the world primarily through our eyes, but the world can be explored just as effectively through other senses. Bats hold unquestionable supremacy in the aerial world of sound. Typically, the sounds that are produced by bats range between 20 kHz and 100 kHz. Most musical notes have a frequency of about 0.5 kHz, which demonstrates that bat calls can be as much as 200 times higher than what human ears are able to detect.

As well as catching flying insects directly in their mouths, bats also use the membranes of their wings and tail almost like a baseball glove. They can deftly catch the insects with their membranes and then pass them up toward the mouth.

No other mammals in the Rockies are as misunderstood as bats. They are thought to be mysterious creatures of the night, souls of the dead, and blind, rabid creatures that commonly become tangled in people's hair. In truth, bats are extremely beneficial creatures whose considerable collective hunger for night-flying insects likely results in fewer agricultural pests. Bats seen flying erratically during the day, however, may be infected with rabies. Never pick up or handle any bat that is active during the day.

DESCRIPTION: This medium-sized, brown bat's most noticeable features are its large ears, which are more than half the length of the forearm. The ears are jointed across the forehead at their bases. The median ear edges are doubled and there is a prominent network of blood vessels visible in the extended ears. At rest, the ears are curled and folded so that they almost resemble Bighorn Sheep horns. There is a set of conspicuous facial glands between the eye and nostril on either side of the snout. The belly is lighter brown than the back.

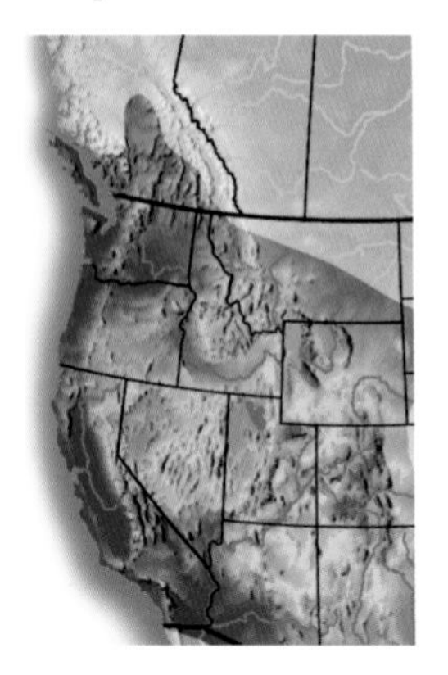

RANGE: This bat ranges through all of western North America south of British Columbia, Montana and South Dakota.

DID YOU KNOW?

Despite derogatory references to bats as "flying rats," they are actually more closely related to primates than they are to mice and other rodents.

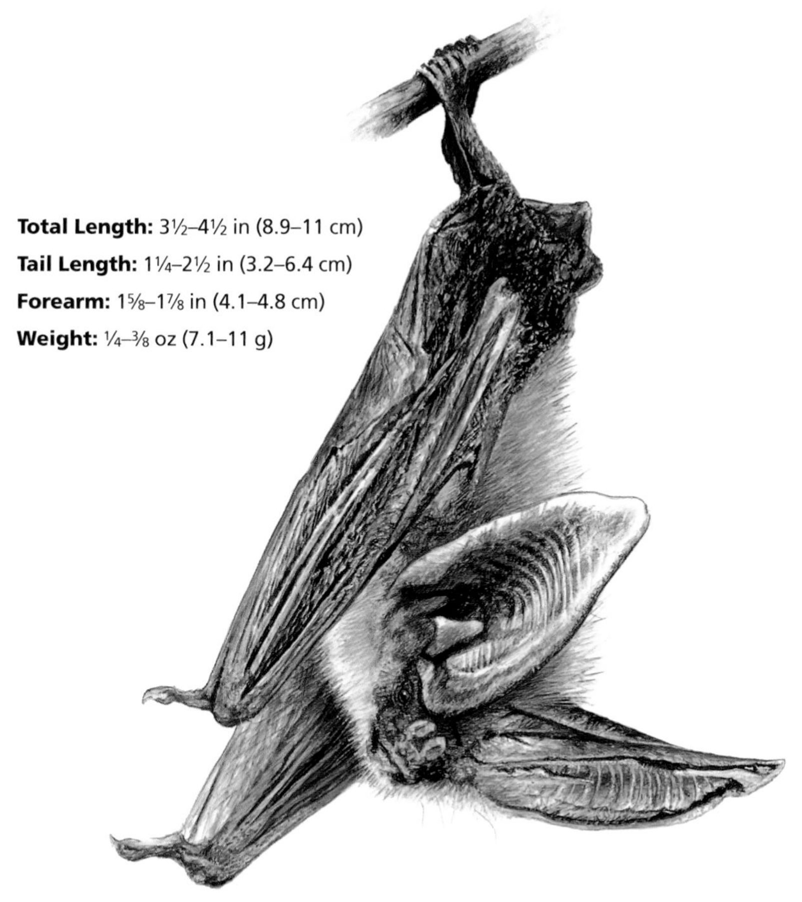

Total Length: 3½–4½ in (8.9–11 cm)

Tail Length: 1¼–2½ in (3.2–6.4 cm)

Forearm: 1⅝–1⅞ in (4.1–4.8 cm)

Weight: ¼–⅜ oz (7.1–11 g)

HABITAT: This bat is found in open places near coniferous forests and in arid areas.

FOOD: These bats emerge quite late in the evening, so they are seldom observed while feeding. They forage along forest edges and are not thought of as gleaners. They principally catch small moths in the air but also readily take beetles, flies and wasps.

DEN: Maternity colonies are found in warm areas of caves and abandoned mines. These colonies are not as large as in other species, and clusters of more than 100 females and young are uncommon. Males in summer tend to be solitary. During the winter hibernation, Townsend's Big-eared Bats tend to move deep into caves, where temperatures are constant.

YOUNG: Mating occurs after ritualized courtship behavior in October and November. The young bats are a quarter of their mother's weight at birth. They are born between May and July. Births appear to be earlier in the year in the southern parts of their range.

SIMILAR SPECIES: Because of its huge ears, this species can only possibly be confused with the Pallid Bat (p. 275), but that bat is much smaller and lacks the lumps on its nose. The Spotted Bat (p. 274) also has huge ears, but its large white spots make it unmistakable.

SHREWS & OPOSSUMS

This grouping, unlike the others in this book, actually encompasses two separate orders of mammals: shrews belong to the insectivore order; opossums are marsupials. Although shrews and opossums are different in many ways, they are united in their classification as the most primitive of the Rockies' mammals.

Shrew Family (Soricidae)

Shrews first appeared back in the times of the Cretaceous dinosaurs, and biologists consider modern shrews to be most similar to the very earliest placental mammals. Many people mistake shrews for very small mice. Shrews don't have a rodent's prominent incisors, however, and they generally have smaller ears and long, slender, pointed snouts.

Pygmy Shrew

Because they are so small, shrews lose heat rapidly to their surroundings. These tiny mammalian furnaces use energy at such a high rate that they may eat three times their own weight in food in a day. Some shrews have a neurotoxic venom in their saliva that enables them to subdue amphibians and mice that outweigh them. Shrews do not hibernate, but their periods of intense food-searching activity, which last 30 to perhaps 45 minutes, are interspersed with hour-long energy conserving periods of deep sleep, during which the body temperature drops.

Of all the Rocky Mountain shrews, only the Common Water Shrew and the Arctic Shrew are reasonably easy to identify visually, as long as you can get a long enough look at them. The other species must be distinguished from one another on the basis of tooth and skull characteristics, distribution and to some extent habitat, although in many cases ranges and even habitat may overlap.

Opossum Family (Didelphidae)

The Virginia Opossum is the only marsupial in North America north of Mexico. Marsupials get their name from the marsupium, or pouch, where newborns are typically carried. Because they usually have no placenta, marsupials bear extremely premature young that range from honey bee to bumble bee size at birth. Once in the marsupium, the young attach to a nipple and continue the rest of their developmental period outside the uterus. There is also a difference in dentition between placental mammals and marsupials: early placental mammals typically have four premolars and three molars; early marsupials have three premolars and four molars. The majority of the world's marsupials live in Australia and New Zealand, with a few in South and Central America.

Virginia Opossum

Masked Shrew

Sorex cinereus

The Masked Shrew may be the most common shrew across much of the Rocky Mountains, but in spite of its wide distribution and abundance, few are ever seen alive. This shrew follows its pointed nose and long whiskers through a world of underbrush and tall grass in both deciduous and coniferous forests. You are more likely to see one dead in spring—starvation in the late winter months claims many—its body to be recycled during the upcoming bursting forth of life.

DESCRIPTION: These medium-sized shrews have dark brown backs, lighter brown sides and pale underparts. The winter coat is paler, and the fur is short and velvety. The Masked Shrew has a long, flexible snout, tiny eyes, small feet and a bicolored tail, which is dark above and light below.

HABITAT: The Masked Shrew favors forests, either coniferous or deciduous, and sometimes areas of tallgrass prairie or shrubby wooded areas.

FOOD: Insects account for the bulk of the diet, but this shrew also eats significant numbers of slugs, snails, young mice, carrion and even some vegetation.

DEN: The nest, located under logs, in debris, between rocks or in burrows, is about 2–4 in (5.1–10 cm) in diameter and looks like a woven grass ball. The nest does not have a central cavity; the shrew simply burrows to the inside.

Total Length: 2¾–4¼ in (7–11 cm)

Tail Length: 1–2 in (2.5–5.1 cm)

Weight: 1⁄16–¼ oz (1.8–7.1 g)

YOUNG: Mating occurs from April to October, and, with a gestation period of about 28 days, a female may have two or three litters a year. The four to eight young are born naked, toothless and blind. Their growth is rapid: eyes and ears open in just over two weeks, and they are weaned by three weeks.

SIMILAR SPECIES: All shrews look very similar. The Preble's Shrew (p. 280) is slightly smaller and is generally found in areas of dry, short grasses or shrubs.

RANGE: The Masked Shrew occurs across most of Alaska and Canada. Its range extends south into northern Washington, through the Rockies and through most of the northeastern U.S.

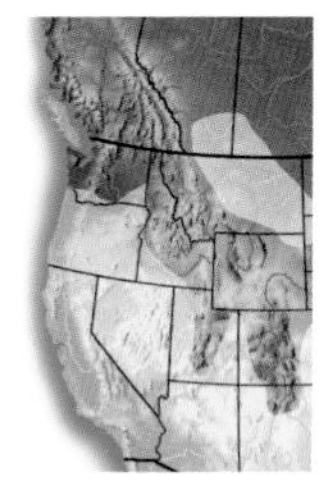

Preble's Shrew

Sorex preblei

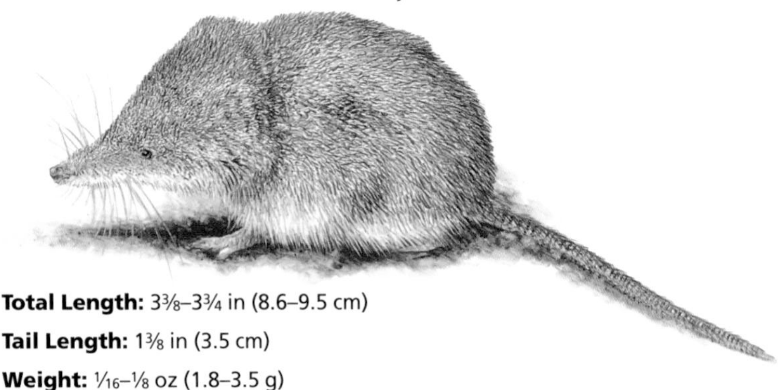

Total Length: 3⅜–3¾ in (8.6–9.5 cm)

Tail Length: 1⅜ in (3.5 cm)

Weight: 1/16–⅛ oz (1.8–3.5 g)

Naturalists have been known to measure their experiences by comparing their wildlife encounters with other naturalists of similar interests. When speaking of the Rockies, they often joust their encounters with such animals as Mountain Lions and Grizzly Bears to validate their experience. A true measure of a naturalist's character, however, may not lie with such sightings as these, but instead with appreciating the under-appreciated. The Preble's Shrew is so rarely seen that any naturalist who comes across it can count him- or herself exceptionally lucky.

DESCRIPTION: The Preble's Shrew has a brownish-gray back that grades to lighter colors on the sides and underside. If you raise the upper lip on the side of the snout, you will see four single-pointed teeth behind the large, lobed first incisor. The third of these unicuspid teeth is not smaller than the fourth.

RANGE: Preble's Shrews are found in extreme southeastern Washington, eastern Oregon and most of Idaho and Montana.

HABITAT: This tiny shrew seems to prefer dry sagebrush desert environments or grasslands with rocky areas. Within the Rocky Mountains, it has been captured in subalpine coniferous forests.

FOOD: The Preble's Shrew is thought to eat mostly invertebrates, such as caterpillars, beetles, crickets, wasps and spiders.

DEN: The den is often found in soft soil, among rocks or under woody debris. The nest chamber is exceedingly small, and the entrance to the burrow is small and indistinct.

YOUNG: Little is known about this shrew's reproduction, but it is probably similar to other shrews. Mating likely occurs from April through July, with females having multiple litters a year.

SIMILAR SPECIES: The Vagrant Shrew (p. 281) and the Dwarf Shrew (p. 283) can be distinguished from the Preble's Shrew by looking at their teeth. Individuals of both those species have a fourth unicuspid that is larger than the third.

Vagrant Shrew

Sorex vagrans

The Vagrant Shrew and the Dusky Shrew (p. 282) may be the most difficult mammals in the Rocky Mountains to distinguish from one another. Even experts have trouble telling whether the two tiny, medial tines on the upper incisors are located near the upper limit of the dark tooth pigment (Vagrant Shrew) or within the pigmented part of the incisor (Dusky Shrew). Naturally, live shrews would never submit gladly to such scrutiny, but at least it is ordinarily only an issue where the two species' ranges overlap in the Rocky Mountains south of the Canadian border.

ALSO CALLED: Wandering Shrew.

DESCRIPTION: The back and sides are pale brown in summer. In winter, the back is slightly darker. The undersides vary from silvery gray to buffy brown. The tail is bicolored: whitish below; pale brown above.

HABITAT: The Vagrant Shrew favors forested areas near water. Sometimes it occurs in moister habitats, such as the edges of mountain brooks with willow banks.

FOOD: The diet comprises a variety of adult and larval insects, earthworms, spiders, snails, slugs, carrion and even some vegetation.

Total Length: 3⅜–4¾ in (8.6–12 cm)

Tail Length: 1⅜–1⅝ in (3.5–4.1 cm)

Weight: 3⁄16–¼ oz (5.3–7.1 g)

DEN: The spherical, grassy nest is usually built in a decayed log. It lacks a central cavity.

YOUNG: Mating begins in March, and litters of two to nine young are born from early April to mid-August. Females likely have more than one litter a year. The young are helpless at birth, and they must feed heavily from their mother to complete their rapid growth. The eyes and ears open in about two weeks, and the young are weaned soon thereafter.

SIMILAR SPECIES: The Dusky Shrew (p. 282) is found over all of the Rocky Mountains and generally prefers moister habitats. The Pygmy Shrew (p. 288) is smaller.

RANGE: The Vagrant Shrew's range extends from central British Columbia south to northern California, Nevada and northern Utah.

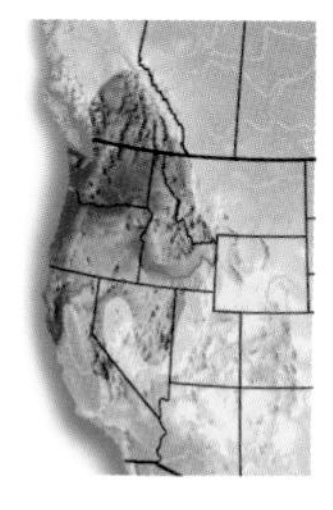

Dusky Shrew

Sorex monticolus

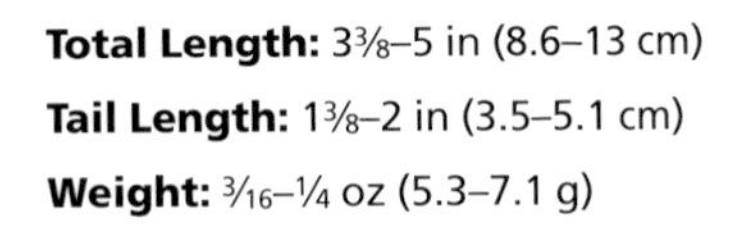

Total Length: 3⅜–5 in (8.6–13 cm)

Tail Length: 1⅜–2 in (3.5–5.1 cm)

Weight: 3/16–¼ oz (5.3–7.1 g)

Given their tiny size, shrews can be remarkably fierce mammals. This observation may surprise most people who have never experienced a shrew up close, but some biologists who have worked with both shrews and bears say they prefer to study bears. The proper identification of shrews almost invariably involves an examination of their teeth, and living shrews, which understandably dislike such close scrutiny, often try to bite the offending fingertips, which is why this exercise is for the serious naturalist only.

ALSO CALLED: Montane Shrew.

DESCRIPTION: This mid-sized shrew has a pale brown back and sides in summer. Its back is slightly darker in winter. The undersides are silvery gray to buffy brown. The bicolored tail is whitish below and the same color as the back above.

RANGE: The Dusky Shrew is found from Alaska east to Manitoba and south into the Cascades and along the Rocky Mountains into Mexico.

HABITAT: The Dusky Shrew can be found in moist alpine meadows and wet sedge meadows, among willows along the edges of mountain brooks and in damp coniferous forests with nearby bogs.

FOOD: This shrew eats a variety of adult and larval insects, earthworms, spiders, snails, slugs, carrion and even some vegetation.

DEN: Dusky Shrews usually build their spherical nests in decayed logs. The nest is a simple bundle of grass without a central cavity.

YOUNG: Mating occurs from March to August, during which time a female likely has more than one litter of two to nine young. The young are helpless at birth, and they must nurse heavily from their mother to complete their rapid growth. The eyes and ears open in about two weeks, and they are weaned soon afterward.

SIMILAR SPECIES: The Vagrant Shrew (p. 281) generally prefers drier habitats, and its range only extends a short distance north into Canada. The Pygmy Shrew (p. 288) is generally smaller.

Dwarf Shrew

Sorex nanus

Total Length: 3⅜–4 in (8.6–10 cm)

Tail Length: 1⅜–1¾ in (3.5–4.5 cm)

Weight: 1⁄16–⅛ oz (1.8–3.5 g)

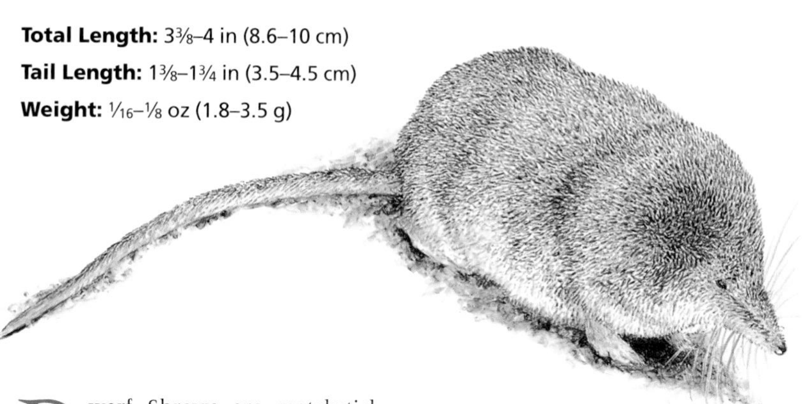

Dwarf Shrews are matchstick-lengthed animals that are found beneath rock rubble in alpine regions and dry brushy slopes. They tend to be more tolerant of drier conditions than other shrews, although they are also found in sedge marshes and ditches. Dwarf Shrews concentrate their activities in areas that naturally serve up meals of convenience. Winds often sweep upslope, carrying insects and spiders up to lofty heights. These drafts weaken along ridges, which is where Dwarf Shrews are most frequently encountered, waiting for their next meal to drop in.

DESCRIPTION: This brownish shrew is sometimes clove brown on its back and gray on the underside. The tail is dark above and light below, but there is no abrupt division between the upper and lower colors. If you lift the upper lip and look at the four unicuspid teeth immediately behind the incisor, the third unicuspid will be distinctly smaller than the fourth.

HABITAT: This shrew is primarily a dweller of montane regions of the eastern front ranges, but it seems to be able to tolerate a range of ecological conditions.

FOOD: Larval insects and other soft-bodied invertebrates are frequent food items. Immature shrews may eat mostly vegetation and spiders.

DEN: The entrance to the burrow is small and indistinct, often no larger than the diameter of a pen. The den is often found in soft soil, in moss or under a log.

YOUNG: Mating likely occurs from April through August, with females having multiple litters a year. Gestation is about 28 days, and litter size is three to seven young.

SIMILAR SPECIES: All other shrews in the Rockies are larger. The Merriam's Shrew (p. 287) and the Masked Shrew (p. 279) have bicolored tails.

RANGE: The Dwarf Shrew ranges from the Alberta-Montana border south to New Mexico.

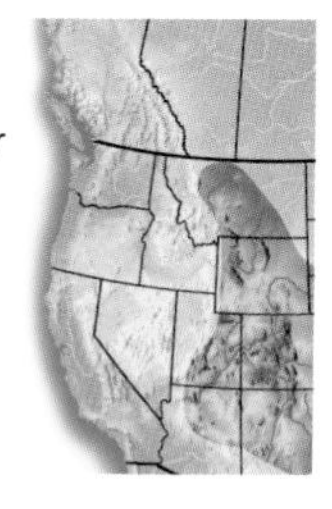

Common Water Shrew

Sorex palustris

Even the shrewdest people would agreed that, by most standards, the majority of shrews in the Rocky Mountains have few distinguishing characteristics. The Common Water Shrew, however, is an exception in the Rockies' shrewdom—this finger-sized heavyweight is so unusual in its habits that it deserves a celebratory status. While other shrews prefer to wreak terror on the small vertebrates and invertebrates roaming on land, the Common Water Shrew literally takes the plunge to feed upon aquatic prey.

In its astonishing foraging dives, this fierce predator swims after its prey, ably seizing insect nymphs, sticklebacks and other small fish and dragging them to land, where they are quickly consumed. The Common Water Shrew is aided in its aquatic pursuits by small hairs on the hindfeet that effectively act as flippers, thereby providing this shrew with the paddle power it needs to swim down prey. Once it is out of the water, the shrew's fringed feet serve as combs with which to brush water droplets out of the fur.

Perhaps the easiest of all shrews to observe, the Common Water Shrew commonly occurs beneath overhangs or ice shelves along flowing waters, particularly small creeks and backwaters. If you are walking along these shorelines, it is not unusual to see a small black bundle rocket from beneath the overhang into the water. The motion at first suggests a frog, but the Common Water Shrew tends to enter the water with more finesse, hardly producing a splash. Often, the shrew first runs across the surface of the water a bit before diving in. Some voles and mice are also scared into or across waters in this way, but even at a quick glance you can distinguish this shrew from those rodents by its smaller size and velvety black color.

DESCRIPTION: The largest long-tailed shrew in the Rockies, the Common Water Shrew has a velvety black back and contrasting light brown or silver underparts. The third and fourth toes of the hindfeet are slightly webbed, and a stiff fringe of hairs around the hindfeet aid in swimming. Males tend to be somewhat larger than females.

HABITAT: This shrew can be found alongside flowing streams with undercut, root-entwined banks, in sphagnum moss on the shores of lakes and occasionally in nearly dry streambeds or tundra regions.

RANGE: This transcontinental shrew ranges from southern Alaska to Labrador and south along the Cascades and Sierra Nevada to California, along the Rocky Mountains to New Mexico and along the Appalachians almost to Georgia.

DID YOU KNOW?

When Common Water Shrews dive into the water, air trapped in their fur transforms them into sleek, silvery torpedoes. To return to the surface they simply stop swimming; the buoyancy of the air pops them back up like a cork.

Total Length: 5½–6¾ in (14–17 cm)

Tail Length: 2⅜–3⅜ in (6–8.6 cm)

Weight: 5⁄16–11⁄16 oz (8.9–19 g)

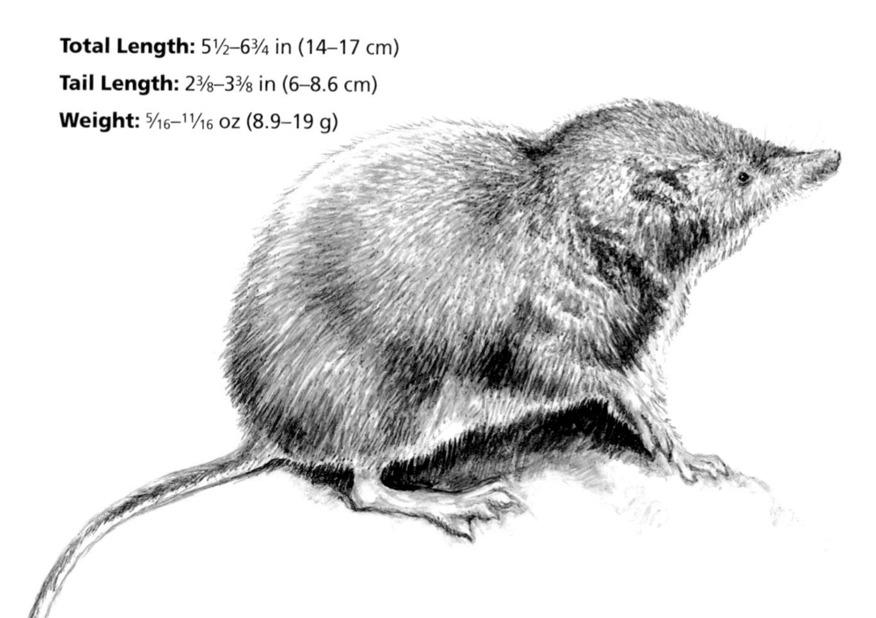

FOOD: Aquatic insects, spiders, snails, other invertebrates and small fish form the bulk of the diet. With true shrew frenzy, these scrappy water lovers may even attack fish that are half as large as themselves.

DEN: This shrew dens in a shallow burrow in root-entwined banks, in sphagnum moss shorelines, or even in the wood debris of beaver lodges. The nest is a spherical mound of dry vegetation, such as twigs, leaves and sedges, that is about 4 in (10 cm) in diameter.

YOUNG: The Common Water Shrew breeds from February until late summer, and females have multiple litters each year. Females born early in the year usually have their first litter in that same year. Litters vary in size from five to eight young, and, as with other shrews, the young grow rapidly and are on their own in a few weeks.

SIMILAR SPECIES: No other Rocky Mountain shrew has the large size (for a shrew) and the velvety black fur of the Common Water Shrew.

bounding trail

Arctic Shrew

Sorex arcticus

Total Length: 4–4¾ in (10–12 cm)

Tail Length: 1½–1¾ in (3.8–4.5 cm)

Weight: 3⁄16–½ oz (5.3–14 g)

Arctic Shrews may be the most handsome of all North American shrews. Although many weasels and hares change color seasonally, it is quite an unusual trait for a shrew. Not only is the Arctic Shrew's winter coat longer and denser than its summer coat, it is also more vibrant, with a coal black back, brown sides and a snowy white belly. The spring molt changes this dress slightly, with the fur being replaced first on the forequarters and progressing to the rear. The full summer coat is less striking, with a brown back and gray underparts.

ALSO CALLED: Saddle-backed Shrew.

DESCRIPTION: The tricolored body of this stocky shrew makes it easily recognizable: the back is chocolate brown in summer and glossy black in winter; the sides are gray-brown year-round; the undersides are ashy gray in summer and silver white in winter. The tail is cinnamon colored year-round. Females are usually slightly larger than males.

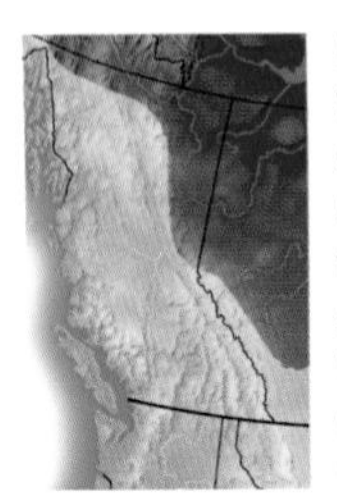

RANGE: The Arctic Shrew is found from the southeastern Yukon across central Canada to Newfoundland and south to Minnesota, Wisconsin and parts of North Dakota and Michigan.

HABITAT: This shrew typically inhabits moist areas of the boreal forest or along its edges. Outside forested regions, the Arctic Shrew takes to open areas, dried-out sloughs and streamside habitats among shrubs.

FOOD: The Arctic Shrew feeds primarily on both larval and adult insects, including caterpillars, centipedes and beetles, but earthworms, snails, slugs and carrion often make up a significant portion of the diet.

DEN: The spherical, grassy nest, 2½–4 in (6.4–10 cm) in diameter, is built in a small pocket in or under logs, under debris or in rock crevices. Unlike the grass nests of many other mammals, but like most other shrews, the Arctic Shrew's nest lacks an interior cavity, even for newborn shrews—the shrews simply burrow their way into the nest.

YOUNG: Breeding takes place between May and August, and females generally have two litters of 4 to 10 young in a season. Females born early in the year may have their first litter in late summer of that same year, but most females do not breed until the next year.

SIMILAR SPECIES: The tricolored pelage of the Arctic Shrew—a dark back, lighter sides and a still lighter belly—best distinguishes it from other shrews. It also tends to be heavier and stockier than most other shrews.

Merriam's Shrew

Sorex merriami

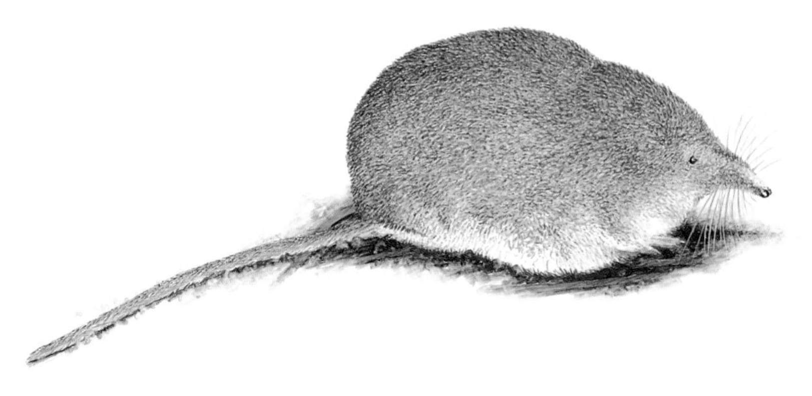

Total Length: 3⅜–4¼ in (8.6–11 cm)

Tail Length: 1¼–1⅝ in (3.2–4.1 cm)

Weight: ⅛–¼ oz (3.5–7.1 g)

When the remains of Merriam's Shrews were found in pottery jars during archeological investigations at Mesa Verde National Park in Colorado, the researchers concluded that the animals were collected intentionally by the Native American inhabitants. Although there is no full explanation to this unusual discovery, it seems unlikely that the shrews were to be eaten, because they are exceedingly small and smelly. The smell of Merriam's Shrews is particularly bad, and their noxious odor might have protected the stored food from rodents.

DESCRIPTION: This shrew has grayish or brownish-gray upperparts and whitish underparts and feet. In winter it is brighter in appearance. The tail, although sparsely furred, is bicolored. The males have very large flank glands. If you lift the upper lip and view the four unicuspid teeth behind the upper incisor, they appear to be crowded together. The second one is the largest, and the third is larger than the fourth.

HABITAT: This shrew inhabits sagebrush flats, deserts, semi-deserts and sometimes dry grasslands.

FOOD: The Merriam's Shrew is thought to eat mostly insects, including beetles, crickets, wasps and caterpillars. Spiders are likely another seasonally common food source.

DEN: Merriam's Shrews make typical shrew nests, often under logs or in soft soil.

YOUNG: Mating occurs from April through July, with females having multiple litters of typically four to seven young in a year.

SIMILAR SPECIES: The Dwarf Shrew (p. 283) is smaller and has a uniform tail color.

RANGE: The Merriam's Shrew has been found from Washington to North Dakota and south to New Mexico and Arizona in appropriate habitat.

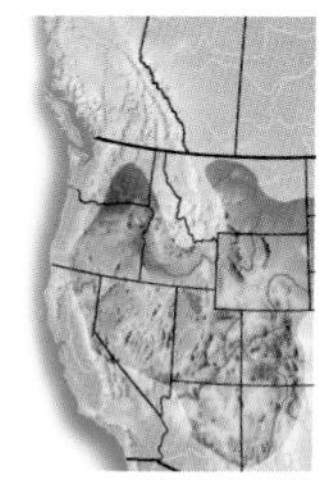

Pygmy Shrew

Sorex hoyi

Total Length: 2⅛–2⅜ in (5.4–6 cm)

Tail Length: 1–1¼ in (2.5–3.2 cm)

Weight: 1/16–¼ oz (1.8–7.1 g)

Weighing no more than a penny, the Pigmy Shrew represents one of the furthest degrees of miniaturization in mammals. In North America, only the Dwarf Shrew (p. 283), which does not occur in the Canadian Rocky Mountains, is smaller. In spite of its size, the Pygmy Shrew is every bit as voracious as other shrews; one studied female ate about three times her body weight each day for 10 days. The Pygmy Shrew may also be one of the rarest shrews in North America.

DESCRIPTION: This tiny shrew is primarily reddish to grayish brown. The color grades from darkest on the back to somewhat lighter underneath. It is usually grayer in winter. The third and the fifth unicuspid teeth are so reduced in size that they may go unnoticed.

HABITAT: The Pygmy Shrew lives in a variety of different habitats, both moist and dry and both forested and open, including deep spruce woods, sphagnum bogs, grassy or brushy areas, cattails and rocky slopes.

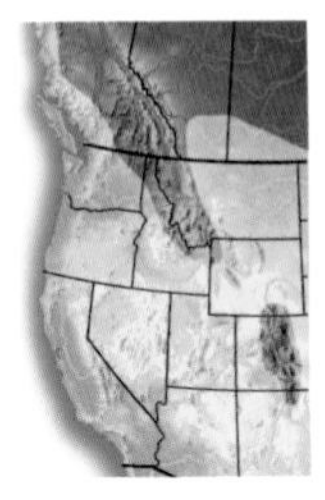

RANGE: The Pygmy Shrew occurs from Alaska east to Newfoundland and south to Colorado, the Appalachians and New England.

FOOD: These shrews feed primarily on larval and adult insects, but earthworms, snails, slugs and carrion often make up a significant portion of the diet.

DEN: The spherical, grassy nest, which is 2½–4 in (6.4–10 cm) in diameter, may be found in a small pocket in or under a log, under debris or in rock crevices. Unlike the nests of many other mammals, there is no rounded cavity inside this grassy ball; instead, the shrew simply burrows its way in among the grass.

YOUNG: Breeding takes place from May until August, and 4 to 10 young are born in June, July or August. Females generally have only one litter a year. Young born early in the year may have a late-summer litter, but most females do not mate until the following year.

SIMILAR SPECIES: The Arctic Shrew (p. 286) is tricolored: it has a dark back, lighter sides and a still lighter belly. Only the Dwarf Shrew (p. 283) is smaller. The Common Water Shrew (p. 284) is much larger and darker.

Desert Shrew

Notiosorex crawfordi

Total Length: 3–3⅝ in (7.6–9.2 cm)

Tail Length: ⅞–1¼ in (2.2–3.2 cm)

Weight: 1⁄16–⅛ oz (1.8–3.5 g)

Like all shrews, the Desert Shrew is strictly territorial, and it will resolutely defend its home range against all intruders. This aggressive tendency often makes courting season a bit confusing. Influenced by contradictory urges to both kill and to mate, this shrew is fastidious in its mate selection. It is thought that the large flank glands of the Desert Shrew serve an important and fragrant purpose during this selection process.

True to it name, the Desert Shrew inhabits arid environments. Like many other small mammals of the desert, it does not need to drink fresh water. Instead, its urinary system is designed to maximize water retention, and it metabolizes enough water from the food it eats.

DESCRIPTION: The Desert Shrew is a gray or brownish shrew with light gray undersides. Unlike other shrews, its ears are conspicuous. The glands on its flanks are larger than those of any other North American shrew. Unlike other shrews, its tooth pigmentation is orange, not burgundy.

HABITAT: This shrew lives in arid environments, such as deserts or sagebrush flats, especially in areas with prickly pear cactus. Other associated vegetation includes agave, assorted cacti and dwarf shrubs.

FOOD: Desert Shrews feed mainly on insects and other invertebrates. Usually only the soft parts are consumed.

DEN: Very little is known about the habits of this shrew, but it is often found inside woodrat nests or in clumped vegetation at the base of a cactus or shrub.

YOUNG: This shrew probably breeds through spring and summer, and perhaps into fall. Litters are suspected to number three to five young.

SIMILAR SPECIES: Although most shrews are difficult to identify visually, the Desert Shrew is unique because of its conspicuous ears.

RANGE: The Desert Shrew is found from southern California east to Arkansas, and south into Mexico.

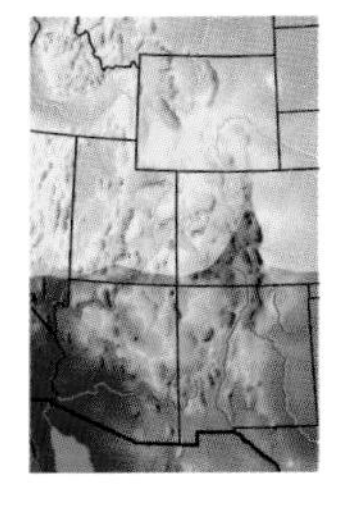

Virginia Opossum

Didelphis virginiana

Among all the mammals of North America, the Virginia Opossum is unique with its prehensile tail, maternal pouch, opposable "thumb" and habit of faking death. Famed by its portrayals in children's literature, the opossum is widely recognized but poorly understood. Few people realize that this animal is a marsupial, and that it is more closely related to the kangaroos and koalas of Australia than to any other mammal native to the U.S. or Canada.

Thanks to the many children's stories, we conjure up images of opossums hanging in trees by their tails. This behavior is not nearly as common as the literature suggests. An opossum's tail is prehensile and strong, but it is unlikely to be used in such a manner unless the animal has slipped or is reaching for something.

The phrase "playing 'possum" is derived from the feigned death scene that is put on by a frightened opossum. If an opossum cannot scare away an intruder through fervent hissing and screeching, it will roll over, dangle its legs, close its eyes, loll its tongue out and drool. Presumably, this death pose is so startling that the opossum will be left alone.

If you are doing much driving through areas where opossums live, it shouldn't be long before you encounter one. Unfortunately, these animals are frequent victims of roadway collisions. Opossums are slow-moving animals that forage at night and find the bounty of road-killed insects and other animals hard to resist. With an abundance of food, opossums may become very fat. Opossums draw upon these fat reserves in winter in the colder parts of their range. Their naked ears and tails do not let them tolerate the cold weather that occurs in most areas of the Rocky Mountains.

DESCRIPTION: This cat-sized, gray mammal has a white face, a long, pointed nose and a long tail. Its ears are black, slightly rounded and nearly hairless. Its tail is rounded, scaly and prehensile. The legs, the base of the tail and patches around the eyes are black. Its overall appearance is grizzled from the mix of white, black and gray hairs.

HABITAT: Moist woodlands or brushy areas near watercourses seem to be favored, but aside from cold habitats, opossums may be found almost anywhere, even in cities.

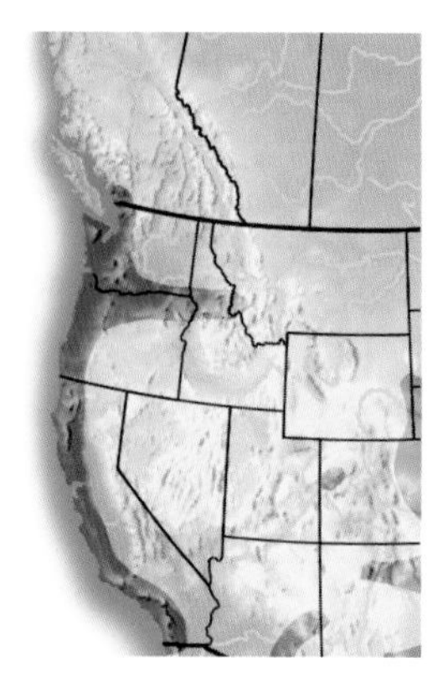

RANGE: The opossum is found in most of the eastern United States and along the entire West Coast as far north as British Columbia. It ranges eastward along the Snake River into Idaho and enters New Mexico from both Arizona and Texas, extending its range into the southern Rocky Mountains.

DID YOU KNOW?

At about the size of a honeybee at birth, an opossum begins life as one of the smallest baby mammals in North America.

Total Length: 27–33 in (69–84 cm)

Tail Length: 12–14 in (30–36 cm)

Weight: 2½–3½ lb (1.1–1.6 kg)

left foreprint

left hindprint

FOOD: A full description of the opossum diet would include almost everything organic. These omnivores eat invertebrates, insects, small mammals and birds, grain, berries, fruit, grass and carrion.

DEN: By day, opossums hide in burrows dug by other mammals, in hollow trees or logs, under buildings or in rock piles. In colder regions, they may remain holed up in a den for days during cold weather, but they do not hibernate.

YOUNG: Up to 25 young may be born in a litter after a gestation period of 12 to 13 days. The young must crawl into the pouch and attach to one of the 9 to 17 nipples if they are to survive. After about three months in the pouch, an average of eight to nine young emerge, weighing about 5½ oz (160 g) each. Females mature sexually when they are six months to a year old.

SIMILAR SPECIES: No other mammal shares the combination of characteristics seen in the opossum. Young, newly emerged from the pouch, might be mistaken for rats, but rats do not have naked, black ears.

Index

Page numbers in **boldface** type refer to the primary, illustrated species accounts.